Urban Forests and Landscape Ecology

Urban Forests and Landscape Ecology

Editors

Manuel Esperon-Rodriguez
Tina Harrison

MDPI • Basel • Beijing • Wuhan • Barcelona • Belgrade • Manchester • Tokyo • Cluj • Tianjin

Editors
Manuel Esperon-Rodriguez
Western Sydney University
Australia

Tina Harrison
Rutgers State University New
Brunswick
USA

Editorial Office
MDPI
St. Alban-Anlage 66
4052 Basel, Switzerland

This is a reprint of articles from the Special Issue published online in the open access journal *Forests* (ISSN 1999-4907) (available at: https://www.mdpi.com/journal/forests/special_issues/Urban_Forests_Landscape).

For citation purposes, cite each article independently as indicated on the article page online and as indicated below:

LastName, A.A.; LastName, B.B.; LastName, C.C. Article Title. *Journal Name* **Year**, *Volume Number*, Page Range.

ISBN 978-3-0365-4577-6 (Hbk)
ISBN 978-3-0365-4578-3 (PDF)

Contents

About the Editors

Manuel Esperon-Rodriguez

Manuel Esperon-Rodriguez (he/him/his) is a Research Fellow in Urban Transformation and Climate Change at the Hawkesbury Institute for the Environment at Western Sydney University. Dr. Esperon did his doctorate of philosophy (Ph.D.) at the National Autonomous University of Mexico (UNAM). His research mainly focused on plant ecophysiological responses to climate change. Then, he moved to Australia in 2015, where he joined Macquarie University as a Postdoctoral Research Fellow and, in 2017, joined the Hawkesbury Institute for the Environment, working on urban ecology and plant eco-physiology. Dr. Esperon is interested in species responses to climatic and environmental changes, and his current areas of research mainly focus on urban ecology, landscape ecology, plant ecology, vulnerability, and climate change. The main goals of his current research are to assess the factors that contribute to the success or failure of urban plantings and the additional benefits that urban greening brings to the cities, such as greater biodiversity, cleaner air, and cooler temperatures and associated benefits to human health and well-being. He is currently a Topic Editor for *Forests*, a Scientific Mentor for *Frontiers for Young Minds*, and review editor for *Frontiers in Plant Science* (Plant Abiotic Stress) and for *Frontiers in Forests and Global Change* (Tropical Forests).

 *forests*

MDPI

Editorial

Urban Forests and Landscape Ecology

Manuel Esperon-Rodriguez [1,*] **and Tina Harrison** [2]

1   Hawkesbury Institute for the Environment, Western Sydney University, Richmond, NSW 2753, Australia
2   Department of Ecology, Evolution, and Natural Resources, Rutgers State University New Brunswick, New Brunswick, NJ 08901, USA; tinaharrison09@gmail.com
*   Correspondence: m.esperon-rodriguez@westernsydney.edu.au

Citation: Esperon-Rodriguez, M.; Harrison, T. Urban Forests and Landscape Ecology. *Forests* **2021**, *12*, 1620. https://doi.org/10.3390/f12121620

Received: 16 November 2021
Accepted: 20 November 2021
Published: 24 November 2021

Publisher's Note: MDPI stays neutral with regard to jurisdictional claims in published maps and institutional affiliations.

Urbanization has become a dominant driver of landscape transformation across the world, with cities representing centers of economic and socio-cultural development. Today, more than 4.2 billion people live in urban areas, which represent ~3 percent of the Earth's land area. By 2050, it is predicted this number will increase to 6.6 billion people (~70% of the predicted global population) [1]. As the human population grows, cities around the globe will continue to expand, increasing the demand for food and services.

Within cities, urban forests provide multiple nature-based solutions as well as other environmental services and socio-economic benefits, which include, among others, heat mitigation, a reduction in air pollutants, the improvement of human health, and social integration [2,3]. Urban forests are also important for coping with psychological stress during events such as the COVID-19 pandemic [4]. Therefore, urban forests are a priority for basic and applied forest research because they are intimately connected with people's physical, cultural, and economic well-being in the urban environment, and can also be important reservoirs of biodiversity.

Therefore, landscape ecology provides a critical perspective for understanding and managing urban forests. Urban forests exist within the matrix of socio-ecological systems and in spatially complex configurations, which affect key processes, including growth, reproduction, and interactions with animal mutualists and commensals, which also link urban forests to surrounding non-forest habitats. Therefore, urbanization's direct impacts on forests are modified (and often compounded) by concurrent, indirect impacts via changes to forests' spatial configurations and surrounding habitats.

To promote a better understanding of urban forests and landscape ecology, this Special Issue in "Urban Forests and Landscape Ecology" compiled research set in urban forests and focused on some spatially explicit processes. Studies presented in this Special Issue are highly interdisciplinary and use a wide range of research approaches. The aim is to increase the number of studies based on rapidly urbanizing areas and increase the knowledge regarding urban forests and landscape ecology.

Conflicts of Interest: The authors declare no conflict of interest.

References

1. UN. United Nations: The world's cities in 2018. In *Department of Economic and Social Affairs: Population Division, World Urbanization Prospects*; United Nations: New York, NY, USA, 2018; pp. 1–34.
2. Livesley, S.; McPherson, E.G.; Calfapietra, C. The urban forest and ecosystem services: Impacts on urban water, heat, and pollution cycles at the tree, street, and city scale. *J. Environ. Qual.* **2016**, *45*, 119–124. [CrossRef] [PubMed]
3. Keeler, B.L.; Hamel, P.; McPhearson, T.; Hamann, M.H.; Donahue, M.L.; Prado, K.A.M.; Arkema, K.K.; Bratman, G.N.; Brauman, K.A.; Finlay, J.C. Social-ecological and technological factors moderate the value of urban nature. *Nat. Sustain.* **2019**, *2*, 29. [CrossRef]
4. Weinbrenner, H.; Breithut, J.; Hebermehl, W.; Kaufmann, A.; Klinger, T.; Palm, T.; Wirth, K. "The forest has become our new living room"—The critical importance of urban forests during the COVID-19 pandemic. *Front. For. Glob. Chang.* **2021**, *4*, 68. [CrossRef]

*forests*

Article

Analysis of the Structure and Ecological Function of an Extreme Landscape in a Tropical Region of West Java, Indonesia

Susanti Withaningsih [1,2,3,*], Parikesit Parikesit [1,2,3], Annas Dwitri Malik [1,3] and Muthi'ah Aini Rahmi [1,3]

[1] Master Programme in Sustainability Science, Graduate School, Universitas Padjadjaran, Bandung 40132, Indonesia; parikesit@unpad.ac.id (P.P.); annas.dm27@gmail.com (A.D.M.); muthi.rahmi@gmail.com (M.A.R.)

[2] Department Biology, Faculty of Mathematics and Natural Sciences Natural Science, Universitas Padjadjaran, Sumedang 45363, Indonesia

[3] Center for Environment and Sustainability Science, Universitas Padjadjaran, Bandung 40132, Indonesia

[*] Correspondence: susanti.withaningsih@unpad.ac.id; Tel.: +62-22-2502176

Abstract: An extreme landscape is a spatially heterogeneous area with unusual topography that is prone to natural disasters but still exhibits interrelated structures and functions. One of the important functions of an extreme landscape is its ecological function. This study aimed to determine the structure and reveal the ecological functions of an extreme landscape in a tropical region of West Java, with special reference to Rongga Sub-district. The method used was a combination of remote sensing techniques and geographic information systems, which were required to process, analyze, and interpret Landsat 8 OLI/TIRS data. The landscape structure was quantified by landscape metrics, after which an analysis of ecological functions was carried out based on the constituent elements of the landscape. The results showed that the landscape structure of Rongga Sub-district consists of various elements of agroforestry land, open fields, settlements, shrubs, plantations, and rainfed and irrigated rice fields. Additionally, secondary forest land acted as a landscape matrix where rivers crossed as natural corridors. The amount of each element varied; agroforestry land had the highest value, indicating that this element showed a high degree of human intervention. Each patch was adjacent to other patch types, and the landscape diversity was quite high. The extreme topography of Rongga Sub-district supports the landscape connectivity and consequently the presence of wild animals in this area. Therefore, Rongga Sub-district has an essential ecological function as a refuge for protected animals living in non-conservation areas.

Keywords: ecological function; landscape structure; remote sensing; GIS

1. Introduction

A landscape is a spatially heterogeneous area [1] characterized by a mosaic of patches differing in size, shape, content, and history [2]. In general, landscapes are formed by the interaction between natural (ecological) factors and human factors [3]. As a result of this interaction, landscapes have a variety of visual, cultural, and ecological constructs. These landscape characteristics make an area unique because of the different element patterns in certain landscape types [4].

A landscape shows the same three basic characteristics of all living systems: structure, function, and dynamics. Structure is the spatial relationship between the different ecosystems (or elements) that make up a landscape. Function is the interaction between spatial elements: the flow of energy, materials, and species between ecosystem components and the intrinsic behavior of complex mosaics. Dynamics comprises the evolution and changes in the structure and function of complex mosaics over time [5].

The structure of a landscape has an important influence on its functional characteristics. Any changes in the landscape structure, whether spatial or temporal, affect energy and material flows, the feasibility and habitability of the landscape, ecological stability, and

Citation: Withaningsih, S.; Parikesit, P.; Malik, A.D.; Rahmi, M.A. Analysis of the Structure and Ecological Function of an Extreme Landscape in a Tropical Region of West Java, Indonesia. *Forests* **2022**, *13*, 115. https://doi.org/10.3390/f13010115

Academic Editors: Manuel Esperon-Rodriguez and Tina Harrison

Received: 5 October 2021
Accepted: 11 January 2022
Published: 13 January 2022

Publisher's Note: MDPI stays neutral with regard to jurisdictional claims in published maps and institutional affiliations.

other characteristics [6]. One of the functional characteristics of a landscape is its ecological function. Ecological functions are defined as physical, chemical, and biological processes that play a role in maintaining the balance of natural ecosystems as well as providing life support systems such as water, soil, and air [7]. This function is also related to species interactions that can maintain biogeochemical cycles, support ecosystem productivity, and prevent extinction [8].

The landscape structure can affect various ecological functions, such as nutrient distribution, light intensity, and groundwater saturation, and can indirectly affect species distribution [9,10], especially in areas with extreme conditions. Several studies have shown the importance of topography and patterns occurring in areas with sloping landscapes, including the flow of nutrients and water from top to bottom and its influence on the chemical and physical properties of the soil [11], as well as the interactions between vegetation and light that can cause differences in the tree community composition and the distribution of many tree species [12]. The relationship between the ecological structure and the function of a landscape can also be demonstrated by the presence of landscape elements such as green corridors that can facilitate individual exchange between habitat patches, encourage genetic variation, and reduce population fluctuations [13].

An increase in agricultural, urban, and rural areas, as well as a decrease in forest areas, can hamper regulatory functions (material and energy flows). Climate regulation and gas regulation functions decline, resulting in a loss of value in terms of ecosystem services due to greenhouse gas emissions and climate change. The significant decrease in the number of agricultural and forest areas between 2013 and 2017 in Guizhou Province, China, led to a decline in soil fertility and the escalation of erosion. The regulatory function of forming and protecting soil was badly damaged, and the ability to obtain, store, process, and recycle nitrogen, phosphorus, and other nutrients were also reduced. During the research period, the number of landscape patches on agricultural land and forests continued to increase; the degree of fragmentation increased sharply; the degree of landscape aggregation and adjacency decreased; and the dominant landscape connectivity, namely forest areas, began to decline, causing the biological habitat to be severely damaged [14].

From the abovementioned explanations, it is clear that landscape structure and ecological functions are interrelated. The flow of matter and energy in a landscape depends on the elements that form a pattern in the landscape. Thus, the structure will determine the ecological function of the landscape. Ecological function is essential in a landscape because it affects the processes that occur in the landscape; the interaction between matter and energy forms a system of abiotic and biotic components and leads to the survival of organisms (ecological balance) [15].

This paper is based on fieldwork conducted in a humid region of West Java, i.e., Rongga Sub-district. The surveys aimed to collect data on the structure of an extreme landscape and the presence as well as the distribution of wildlife within the landscape. A partial analysis focused only on wildlife distribution has been published elsewhere [16–20]. A more comprehensive and integrated analysis concerning the structure of an extreme landscape and how this landscape has an important ecological function through the provision of wildlife habitats is necessary. Therefore, this paper elucidates how an extreme landscape could exhibit its ecological functions as an important refuge for protected animals living outside conservation areas.

The topography of Rongga Sub-district is dominated by hilly and mountainous landscapes where many agricultural activities are practiced. The presence of various land uses such as upland cash crop agriculture, rice fields, agroforestry, and production forest makes the landscape heterogeneous. Despite the fact that the landscape is mainly used for production purposes, such as for foods, timber, and livestock forages, the present landscape structure could also be directed for wildlife conservation. Through the combination of landscape capability and integrity concepts, it is expected that the agrobiodiversity condition in the extreme landscape could fulfil the basic needs of the local people and simultaneously conserve local biodiversity. Nevertheless, due to its complex morphology, such an area may have a relatively high soil erosion risk and more severe mass movement that results in landslides. In this regard, Rongga Sub-district can be considered an area with extreme conditions because its landscape exhibits unusual topographical conditions dominated by areas with a slope of more than 45% and is prone to natural disasters such as landslides and soil erosion.

2. Materials and Methods

2.1. Study Area

Rongga Sub-district, with an area of 229 km^2, is a fertile area and has scenic beauty with distinct topographical conditions (hills with varying heights and slopes). All villages in Rongga Sub-district, namely Cicadas, Cibedug, Sukamanah, Bojong, Bojong Salam, Cinengah, Sukaresmi, and Cibitung, have a slope/peak topography (Figures 1 and 2). Based on an analysis of land suitability in the RTRW of Bandung Regency, 2001–2010, Rongga Sub-district is an area with land that is very suitable for annual crops/agroforestry, with rainfall of 2500–3000 mm/year [21].

The areas in Rongga Sub-district have different slope profiles. The area of flat slope in the Sub-district of Rongga is 1700 hectares. A total of 292 ha is in the rather steep category (15%–25%), 507 ha is in the steep category (25%–40%), and 8812 ha is in the very steep category (>40%) [22].

Geologically, West Bandung Regency is an area that has the potential for earthquakes, especially tectonic-type and volcanic earthquakes. Landslides are also natural disasters that often hit the West Bandung Regency area. Landslides can be caused by ground movement due to scouring water from heavy rains. Rongga is a sub-district that has medium–high ground movement potential due to frequent landslide occurrence. In this zone, soil movement can occur if the rainfall level is above normal, while the old soil movement can be active again [23]. Rongga Sub-district is a non-groundwater basin zone with low aquifer productivity; therefore, it is not suitable for development, except for shallow aquifers in valley areas for drinking and household water needs with a maximum extraction of 100 m^3/month per well [22].

Based on demographical data from Rongga Sub-district in 2021, the total population is 60,666 with the largest population in Cibitung Village, as many as 10,232 people, and the smallest in Bojong Salam at 5339 people. The population density in the subdistrict is approximately 264 people/km^2, with a sex ratio value of 106.2. There are 14,549 people in the age group of 0–4 years, 41,434 people are 15–64 years old, and 4683 people are over 65 years old [24].

2.2. Research Method

Physical environment data were required in this study. This information was gathered from national and regional agencies that provide the related data as well as through a literature study. The gathered physical information was extracted into spatial data in order to be included in the landscape analysis. The extracted spatial data included topography, rainfall, soil types, and their derived data. Maps and satellite imageries were required for analyzing the physical conditions and land use–cover of the study area. The collected data and information were analyzed simultaneously in order to characterize the extreme landscape and quantify the elements that make up the landscape structures.

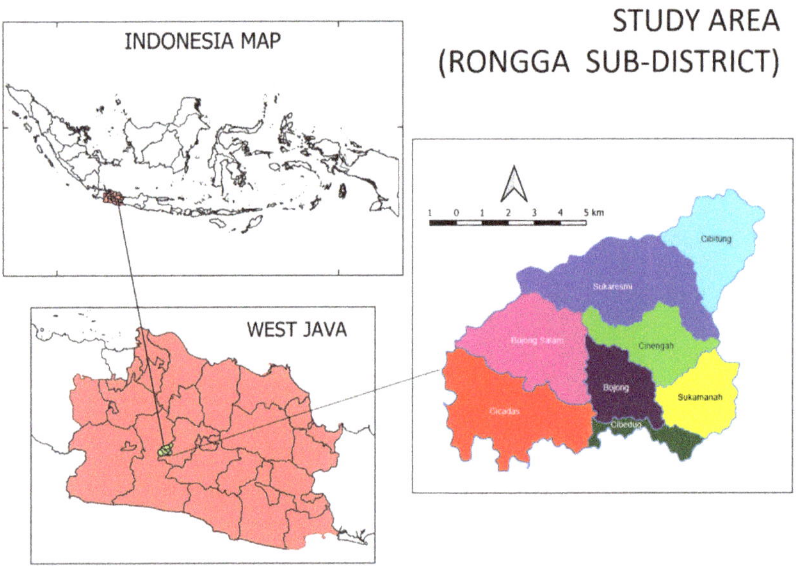

Figure 1. Borders of Rongga Sub-district (Source [25]).

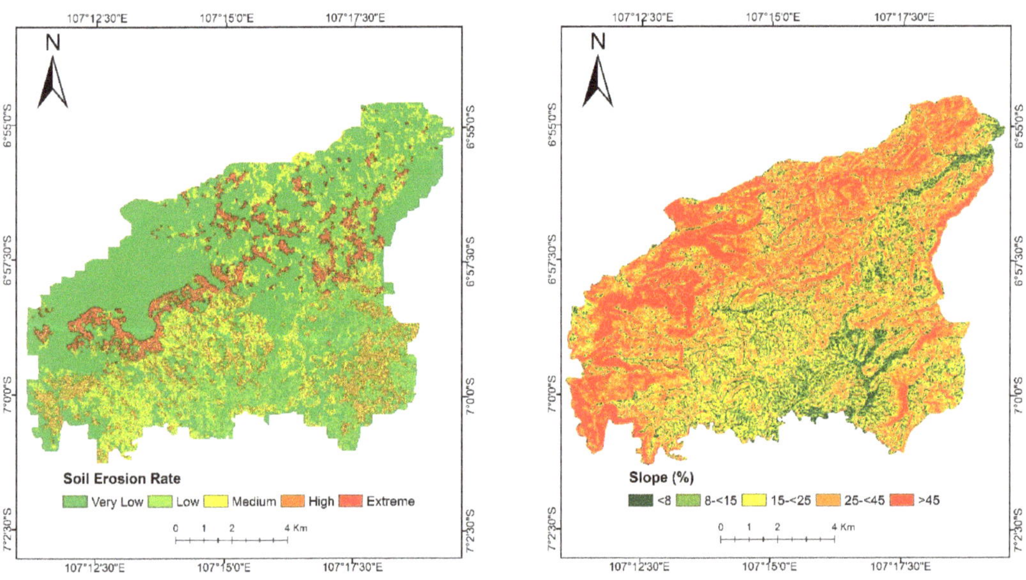

Figure 2. Soil erosion and slope distribution map in Rongga Sub-district.

The method used in this research was a quantitative descriptive method integrating a remote sensing technique and a geographic information system [26]. The remote sensing technique was needed to identify the Landsat 8 OLI/TIRS images and derive information on land use–cover in the study area. The geographic information system method in this study was derived into two approaches: USLE (Universal Soil Loss Equation) modeling, which was widely used to analyze the characteristics of the extreme landscape [27], and the quantification of the landscape structure, which was used to identify the composition, configuration, distribution, and heterogeneity of landscape elements [26].

Analysis at the class level was carried out to identify the types of constituent elements, dominance, number of each constituent element, shapes of the elements, and distribution of the elements. The basic element of a landscape is a patch or corridor (if linear) [28]. A patch is identified by its shape, size, biotic types, number, and configuration. A patch system creates an ecological mosaic to identify a landscape matrix. Thus, the identification of landscape elements means the identification of the patch, matrix, and corridor.

The analysis of ecological function focused on the ecological network formed by different landscape elements and how they were used as habitats to provide ecological services for particular wildlife found in the study area. The GIS technique of least-cost modeling was applied in order to estimate the ecological connectivity of the landscape. An analogy approach using relevant literature was used to conduct an ecological function analysis of the landscape elements related to the built model of ecological networks.

2.3. Land Use–Cover Identification

A Landsat 8 OLI/TIRS image (code LC08_L1TP_122065_20190725_20190801_01_T1), on path/row 122/65, was applied to interpret the land use and land cover of the study area. A medium resolution of 30×30 m of the image data was suitable for the purpose of this study. QGIS 3.10.0 spatial analysis software was used for image analysis and processing. In this study, the Landsat 8 image was adjusted into real-world map coordinates with a projection of the Universal Transverse Mercator (UTM), precisely south of UTM Zone 48, and the World Geodetic System (WGS) 1984 datum configuration.

The first step in the land use and land cover classification was pre-processing the image data, consisting of atmospheric correction and radiometric correction. For atmospheric correction, the dark object subtraction (DOS) technique was used to correct undesirable objects, e.g., shadows, deep water, and dense forests. For radiometric correction, some atmospheric particles such as clouds, ash, and fog were corrected and reduced. To conduct radiometric correction, the digital number (DN) in the satellite imagery was converted to reflectance values.

The next step was generating a band composition of the Landsat 8 satellite image. This band composite feature is useful to make image interpretation easier. Bands 1–7 of the Landsat 8 image were composited or overlaid in order to generate various band combinations that would be used to conduct an image classification. In order to produce a better resolution of the satellite image, a pan-sharpening technique was used on Band 8 of the Landsat imagery. Different band combinations may influence the image output, and each image produced by a typical band combination has its own capacity to be interpreted. A natural color combination of RGB 432 is better to visualize the actual aerial view of the area than other band combinations. However, in order to visualize a better vegetation appearance, a false color band combination of RGB 764 is the most suitable combination as vegetation can reflect the near-infrared (IR) bands much better than other types of land cover.

Image cropping was the next step. A vector boundary of the Rongga Sub-district administrative border collected from the Indonesia Geospatial Agency was used as the area of interest. In addition to the administrative border, geodetic coordinates and a zooming box feature included in the map processing software were also used. The image cropping step is useful to focus on the area that will be analyzed and reduce the memory load while generating the image classification.

In order to generate a land use and land cover map, on-screen interpretation and image classification was selected to derive spatial information from the Landsat 8 image. The image interpretation was assisted by the QGIS 3.10.0 software through a training sample collection and a supervised classification technique. Some vector polygons were drawn to the pixels of the processed image. All visually similar pixels were classified into the same class of objects. The number of sample polygons had to fulfill the requirement of accuracy, with a minimum of three polygons for every object class. Supervised classification is a process that has been conducted previously to classify the visually similar pixels of an image into proposed land use and land cover classes based on a training sample collection. This process was assisted by the QGIS Semi-Automatic Classification Plugin (SCP) to produce a raster image of the land use and land cover of the Rongga Sub-district area in the Temporary Instruction File Format (.tiff).

Lastly, an accuracy test was conducted to identify the errors in the classification process; thus, the percentage of accuracy was identified according to the classification result. The Kappa coefficient was used in this accuracy test considering its universality in involving all aspects of accuracy, i.e., producer's accuracy/omission error and user's accuracy/omission error, which were obtained from the confusion matrix.

2.4. Soil Erosion Hazard Assessment

Soil erosion hazard was used as an indicator to characterize the extreme landscape of Rongga Sub-district, considering its complex morphology and slope profile. Agricultural activities practiced in this area may cause relatively high soil erosion and, if not carefully managed, may worsen the susceptibility to landslides.

The USLE (Universal Soil Loss Equation) model was used to estimate the soil erosion hazard in the area [26]. All of the required data were analyzed using QGIS 3.10.0 software. The USLE is as follows:

$$A = R \times K \times L \times S \times C \times P \tag{1}$$

where A is the amount of eroded soil expressed in tons per hectare per year; R is a rainfall erosivity factor; K is a soil erodibility factor; and L, S, C, and P are factors of slope length, slope steepness, vegetation cover or land cover, and erosion control practices, respectively. In order to quantify rainfall erosivity, many approaches have been proposed, one of which is a rainfall erosivity equation developed by Lenvain [29].

$$R = 2.21P^{1.36} \tag{2}$$

where R is a factor of rainfall erosivity and P is the result of monthly rainfall measurement. This study used the monthly rainfall measured by a rainfall station around the area of Rongga Sub-district. Then, these measured data were interpolated using the isohyet technique or inverse distance weighting (IDW) interpolation in order to generate raster data of the rainfall erosivity of the study area. The erodibility factor (K) describes the sensitivity of the soil to erosion. The following equation was used to identify the erodibility of each soil type in the study area [26]:

$$100K = (2.71 \times M^{1.14}(10-4)(12-OM) + 3.25(s-2) + 2.5(p-3))/7.59 \tag{3}$$

where K is an erodibility factor or impact of soil profile characteristics and soil type, M is the grain size of the soil or (percentage of sand + mud) × (100-percentage clay), OM is the organic matter percentage, s is an index of soil structure, and p is an index of soil permeability.

After the spatial information of soil type distribution was generated, we inserted the result of the erodibility factors into the spatial data of the soil type distribution map by adding the values to new fields in the map attributes; then, the rasterization of the map was conducted. Factors L and S represent the impact of topography on soil erosion, and these factors are expressed as one unit (LS), where L is a factor of slope length, and S is a factor of slope steepness of the area. Slope data were analyzed by conducting a raster analysis with the Slope technique on the 0.27-arcsecond or ~8-m resolution digital terrain model (DTM) of the study area acquired from the Indonesia Geospatial Agency. In this study, LS was estimated by conducting a terrain analysis with the Hydrology technique on the same digital terrain model (DTM) spatial data. This approach has been integrated into the System for Automated Geoscientific Analyses (SAGA). The method selected for this terrain analysis was developed by the authors of [30], who conducted a terrain analysis to generate LS in a larger scale landscape using the availability of digital terrain models. The factor C represents the impact of vegetation coverage on soil erosion and is heavily related to the utilization of soil. Factor P represents the impact of soil conservation practices on soil erosion [31]. In this study, C and P are expressed as one unit considering the variability of land cover classes in the area, and if soil conservation practices were not implemented, the values of P were considered as 1 [32]. To generate the vegetation coverage map of Rongga Sub-district, image interpretation using QGIS 3.10.0 software through a training sample collection and the supervised classification technique were used. The values of C were acquired from the literature study.

Lastly, to produce the soil erosion hazard map, these USLE factors had to be expressed as raster datasets. A raster calculator was applied to calculate and overlay all the factors in the USLE model. In the produced soil erosion distribution map, the study area was divided into grids of ~8 × 8 square meters, where each square grid consisted of the value of estimated soil erosion (ton/ha/year).

2.5. Landscape Structure Analysis

The landscape comprised several elements, and the matrix was selected to identify the land cover that is dominant and interconnected over the majority of the land surface. The class area (CA) provides a quantification of the landscape consisting of a particular type of patch; therefore, CA is a fundamental measure of landscape composition. Matrix units are expressed in hectares [33]. PLAND can show the area of land use of a class or elements in a landscape, expressed in percentages. SHDI refers to the diversity of patches in the area, and SHEI identifies the distribution (regular or irregular) of patches in the area.

2.6. Ecological Network at Landscape Level

The first step to assess ecological connectivity was selecting targeted species. Their habitat requirements, long-distance path, and mobility behavior were used as a basis for connectivity analysis considering that connectivity depends on individual species and cannot be addressed based only on landscape matrices characteristics [34]. Several medium- to large-sized mammals from direct encounters, camera traps, tracks, and other signs were selected as target species to evaluate the ecological corridor within the landscape. The species consisted of *Panthera pardus melas* (Javan leopard), *Prionailurus bengalensis* (Javan leopard cat), *Hystrix javanica* (porcupine), *Manis javanica* (pangolin), *Nycticebus javanicus* (Javan slow loris), and *Paradoxurus hermaphroditus* (Asian palm civet). This study assumed that the ecological corridors respond adequately to the degree to which the extreme landscape facilitates or hinders the mobility of the targeted species.

Defining core habitat patches was the second step. We attempted to define core habitat patches from the species' perspective. This was based on the species distribution, predicted territory, and long-distance path in the study area. Thirteen core habitats were identified for the analysis.

After the targeted species and their core areas were defined, the GIS method of least-cost modeling was applied to describe the ecological network. Least-cost modeling is a quantitative geographic technique that enables the identification of the potential and optimal routes in landscape matrices with a variety of travel costs [35]. Although least-cost modeling was developed for transport geography study, it has been widely applied in the context of ecological networks for multi-targeted species [34,36,37]. In this study, least-cost modeling was applied to estimate cost values referring to the degree of difficulty for the targeted species to access each landscape matrix and identify the potential paths with the lowest cost for the targeted species to move from one core area to another.

For GIS processing, two layers of spatial information are required: the vector data of the 13 core areas and a map of resistance values of the landscape matrices in the form of raster data. Resistance surface represents the degree of friction of a landscape matrix that enables or impedes the movement of the target species [34,35]. We used slope and land use maps to produce a resistance value map through the weighted overlay technique and rasterized the output at 50-m map resolution. The sum of influences was set to equal

(50:50). Deciding on the resistance values can be a difficult step in least-cost modeling. Due to a shortage of empirical data, we assigned the resistance values of the landscape matrices by consulting many experts on the concerned species' tendency to move into each point of the slope matrix and landscape matrix [34,38]. We set a range of resistance values from 1 to 100 in this study. The resistance values for slope and land use are shown in Tables 1 and 2, respectively.

Table 1. The slope resistance values.

Slope Gradient (%)	Resistance
0–13	10
13–23	20
23–31	30
31–40	40
40–49	50
49–58	60
58–68	70
68–81	80
81–100	90

Table 2. The resistance values for land uses.

Land Use	Resistance
Settlement	100
Irrigated Rice Field	50
Rainfed Rice Field	50
Dry Land Forests	1
Plantations	5
Agroforestry	5
Shrubs	10
Open Fields	40
River	90
Access Road	80

The spatial analysis tool CostDistance provided by ArcMap 10.6.1 was used to estimate the cumulative resistance of the overlaid matrices, defined as the cost of radial displacement for the target species from each core area. The gradient of cost values represents the degree of difficulty for the target species to access each territorial point from the core areas [34]. Next, the CostPath spatial analysis tool was applied to compute least-cost paths connecting all of the core areas. These networks were defined as the ecological corridors with the highest permeability and the lowest inhibition of landscape matrix among the core areas. The least-cost path is the path with the most potential to reduce the cost of mobility of a target species to access a particular territorial point of the core areas, rather than a functional expression of the dispersal process of the target species [39].

2.7. Ecological Function Analysis

Ecological function analysis was carried out using literature reviews. The quantified elements that make up the landscape structure of Rongga Sub-district were then studied in terms of their function on some ecological aspects. An analogy approach with many relevant studies was conducted in order to complete the ecological function analysis of the landscape elements in Rongga Sub-district. Most of the literature used to assist the analysis was based on research that has been previously conducted in Rongga Sub-district [16,17,40,41].

3. Results and Discussion

3.1. Typical Characteristics of the Extreme Landscape

A slope analysis and the USLE soil erosion model were used to confirm the extreme landscape characteristics of Rongga Sub-district. The hilly and mountainous landscape of Rongga Sub-district was divided into five slope classes, expressed in percentages, where slope gradients of <8%, 8%–15%, 15%–25%, 25%–45%, and >45% are considered flat, gentle, moderate, steep, and extremely steep, respectively [29]. The slope distribution of Rongga Sub-district's landscape is shown in Figure 2.

Rongga Sub-district is dominated by steep slopes (25%–25%), covering 4505.37 hectares of the area, followed by moderate slopes (15%–25%), covering 2701.14 hectares of the total area, and extremely steep slopes (>45%), covering 2442.73 of the total area. Meanwhile, the area which was considered a flat slope (0%–8%) covered 708.77 hectares of the total area, indicating this category as the slope category with the smallest area in the Rongga Sub-district landscape. These results also indicate that Rongga Sub-district's landscape is dominated by steep to extremely steep slopes, with gradients over 25% covering most of the area, as shown in Table 3. Agricultural practice on steep slopes, however, is very susceptible to hydrogeological instability [42]. Furthermore, the lack of maintenance of the cultivated land, extreme rainfall erosivity, very weak soil erodibility, and intensive contouring and tillage practices in the area, may worsen the susceptibility to soil erosion [43].

Table 3. Slope classes in the Rongga Sub-district landscape.

Slope Gradient (%)	Level	Area (ha)
0–8	Flat	708.76
8–15	Gentle	1348.38
15–25	Moderate	2701.14
25–45	Steep	4505.36
>45	Extremely steep	2442.73

An erosion risk assessment for the Rongga Sub-district landscape was performed by overlaying five USLE factors using the raster calculator QGIS spatial analyst. The process generated a soil erosion intensity map of the area. The map expresses the intensity of five classes of soil erosion in tons per hectare per year, i.e., very weak (<16), weak (16–60), moderate (60–180), strong (180–480), and extreme (>480). The soil erosion distribution map (Figure 2) shows that the average annual soil loss in the Rongga Sub-district landscape is 103.03 ton/ha/year, and the maximum value of soil erosion is approximately 3195.28 ton/ha/year, which occurred in the area which was not covered by vegetation or was considered as bare lands and extremely steep slopes.

Table 4 show the area of estimated soil erosion occurrence for every intensity class. The estimation of soil erosion area showed that approximately 6816.40 ha (about 60%) of the Rongga Sub-district landscape indicated very weak soil erosion and thus very low erosion risk. Meanwhile, 4592.11 ha (about 40%) of the total Rongga Sub-district area shows a high soil erosion risk. The second most extensive area comprised moderate-intensity soil loss and consequently moderate erosion risk, covering 1546.96 ha (about 14% of the total area). Weak-intensity or low soil erosion risk was in third place, covering 1504.93 ha (13%), followed by strong-intensity or high soil erosion risk, which covered 988.68 ha or about 9% of the total area. Lastly, the extremely high soil erosion risk area was the smallest at 551.53 ha or about 5% of the total area. These results show that the area of very low soil erosion risk is more extensive than that of higher classes of soil erosion risk. However, the area with steep slopes (25%–45%) had the highest average of estimated soil erosion potential at around 116.89 ton/ha/year. Moderate slopes (15%–25%) were next, with an estimated soil erosion average of 114.92 ton/ha/year. Ranking third was gentle slopes (8%–15%), with an estimated soil erosion average of 108.11 ton/ha/year. The estimated soil erosion average value of flat slopes (0%–8%) was not the lowest, being higher (99.68 ton/ha/year) than that of the extremely steep slope area (>45%), which was 81.54 ton/ha/year. These results imply that the soil erosion value shows an increasing trend as the slope gradient increases. However, this only occurred in the 0%–45% slope gradients as the extremely steep slope level (>45%) showed the lowest estimated soil erosion average, as shown in Table 5.

Table 4. Classification of soil loss rate in the Rongga Sub-district landscape.

Soil Loss (ton/ha/year)	Intensity	Area (ha)	Area (%)
<16	Very weak	6816.40	60%
16–60	Weak	1504.93	13%
60–180	Moderate	1546.96	14%
180–480	Strong	988.68	9%
>480	Extreme	551.54	5%

Table 5. Estimated soil erosion average at every level of slope gradient.

Slope Gradient (%)	Level	Soil Erosion Average (ton/ha/year)
0–8	Flat	99.67
8–15	Gentle	108.11
15–25	Moderate	114.92
25–45	Steep	116.88
>45	Extremely steep	81.54

One study showed a positive correlation between the soil erosion average and slope gradient, i.e., the steeper the slope, the higher the estimated soil erosion average [44]. Runoff affected by tillage management may increase soil erosion on longer and steeper slopes [45]. Many other researchers also found that the soil erosion rate shows an exponentially increasing trend with an increasing slope gradient [46].

3.2. Structural Analysis at the Land Cover Type Level

3.2.1. Landscape Elements (Composition) and Dominance

Fragstat 4.2.1 was used to determine the landscape composition and dominance based on the class area (CA) and percentage of landscape (PLAND), and the analysis results are shown in Table 6.

Table 6. Calculation results of CA and PLAND metrics at the class level.

No.	Type/Element	CA (ha)	PLAND (%)
1	Agroforestry	1527.80	13.06
2	Secondary Forest	3498.17	29.88
3	Open Fields	2053.69	17.54
4	Plantations	1136.95	9.712
5	Settlements	742.10	6.34
6	Irrigated Rice Fields	509.87	4.35
7	Rainfed Rice Fields	902.52	7.71
8	Shrubs	1326.32	11.33
9	Rivers	8.89	0.07

Abbreviations: CA, class area; PLAND, percentage of landscape.

Table 1 and Figure 3 show that the landscape in Rongga Sub-district consists of nine elements, namely agroforestry, secondary forests, open fields, rainfed rice fields, plantations, shrubs, irrigated rice fields, settlements, and rivers. Most of the elements were covered with different (natural as well as artificial) vegetation types. The values of CA and PLAND varied, with the highest value shown by the dry land forest patches with an area of 3498.165 ha, i.e., 29.8% of the entire landscape area. The open fields were next, covering an area of 2053.6875 ha, or 17.5% of the landscape, and the third was agroforestry at 13.05%, with an area of 1528.74 ha. A high CA value indicates that the landscape element is the dominant element. A matrix is the most extensive landscape element and has a dominant role in the overall landscape function [28,47]. This suggests that the secondary forest was the matrix in the landscape of Rongga Sub-district.

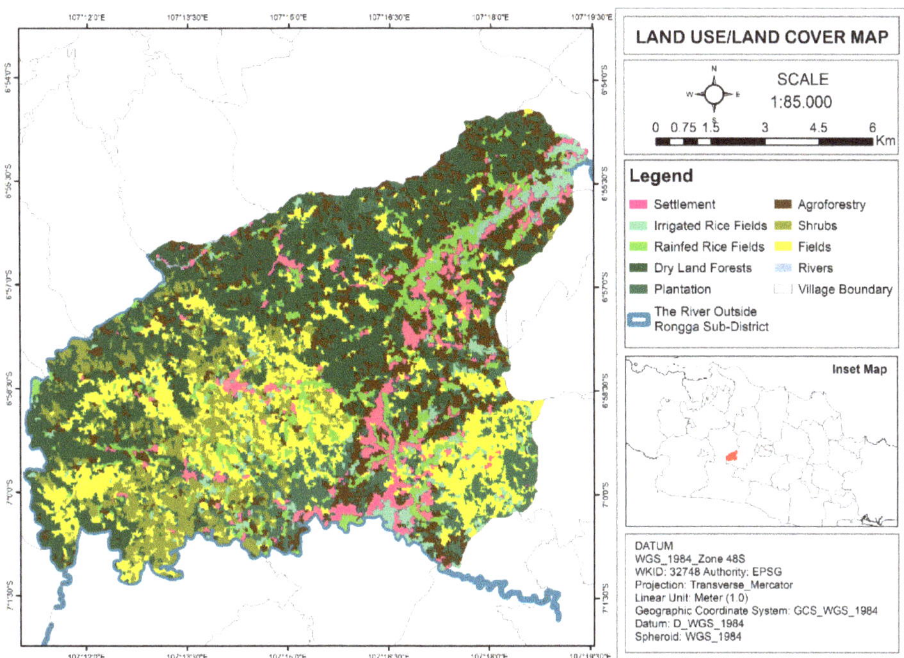

Figure 3. Land use/land cover of Rongga Sub-district.

Forests are an important functional component of a landscape [48]. Secondary forests play a role in environmental protection, wildlife conservation, culture, and economic function. The existence and maintenance of forest ecosystems have a positive impact in terms of improving the quality of the environment and the quality of life of the community [49]. Forests play an important role in soil and water conservation, carbon sequestration and oxygen release, biodiversity conservation, recreation, nutrient accumulation, air filtration, and commodity production [50].

Landscapes dominated by forest matrix patches have high connectivity [51]. The composition of the landscape matrix of the surrounding forest fragments is vital for the survival of some animal species because it offers a structure that helps animals move between fragments and other foraging areas [52].

3.2.2. The Number of Constituent Elements

The elements in the landscape varied in number, as indicated by the number of patches (NP). The results of the analysis using Fragstat 4.2.1 are shown in Table 7.

The number of patches (NP) can be used to measure the spatial heterogeneity of an entire landscape. This is because the NP can measure the quantity of each element that makes up the landscape. The NP is a direct indicator of the level of fragmentation [53]. As shown in Table 7, agroforestry patches were the most numerous in the landscape of Rongga Sub-district with 546 patches, followed by open fields with 429 patches and plantations with 368 patches. This suggests that these three landscape elements experienced severe fragmentation, which resulted in declining connectivity within and between the present main landscape elements in the study area.

Table 7. Number of patches (NP) of each landscape element in the study area.

No.	Type/Element	NP
1	Agroforestry	562
2	Secondary Forest	298
3	Open Field	429
4	Plantation	368
5	Settlement	203
6	Irrigated Rice Fields	245
7	Rainfed Rice Fields	305
8	Shrubs	314
9	Rivers	10

The quantification results showed that the NP of agroforestry was the highest, which is in accordance with this element having the highest CA value. The NP of plantations was the third-highest, but the class area of this type was quite low (fifth out of the nine landscape elements). This showed that plantations were more fragmented compared to other landscape types [54].

The lowest number of patches was obtained for rivers. This was due to the low extent of the river area covered in the study area. In this study, the 1:75,000 scale map created with the Landsat 8 TIRS/OLI satellite data in Rongga Sub-district was not able to show the river areas due to limited pixels and resolution. River patches appeared too narrow and small, so they were less visible on the raster data.

3.2.3. Element Shapes

Elements' patch shapes are an important parameter for describing the landscape structure. The metrics used to quantify this parameter were the landscape shape index (LSI), total edge (TE), and edge density (ED). The results of the analysis using Fragstat 4.2.1 with the LSI, TE, and ED metrics are shown in Table 8.

Table 8. LSI, TE, and ED metrics of each landscape type in the study area.

No.	Type/Element	LSI	TE (m)	ED (m/ha)
1	Agroforestry	39.97	613,320	52.39
2	Secondary Forest	32.34	738,075	63.05
3	Open Fields	35.94	643,740	54.99
4	Plantations	29.91	395,820	33.81
5	Settlements	22.07	237,330	20.27
6	Irrigated Rice Fields	23.05	199,770	17.06
7	Rainfed Rice Fields	27.72	328,965	28.11
8	Shrubs	30.36	435,210	37.17
9	Rivers	5.1	4710	0.40

Abbreviations: LSI, landscape shape index; TE, total edge; ED, edge density.

Based on the LSI results, the landscape elements were arranged from highest to lowest as follows: agroforestry > open fields > secondary forest > shrubs > plantations > rainfed rice fields > irrigated rice fields > settlements > rivers.

Agroforestry had the highest LSI value of 39.97. The value of LSI = 1 occurs when the landscape consists of a square plot or when it is the most compact (almost square) of an appropriate type; the LSI value continues to increase without limit when the type of patch becomes less compact and irregular [55]. An LSI value close to 0 indicates that the landscape has a simple form with high aggregation [56]. Values further from 0 show that the landscape has a complex shape with scattered plots that are segregated. This means that the agroforestry patches had an irregular and complex shape. The lowest LSI value was obtained for a river patch, indicating that this patch was simpler and almost elongated in form and was the most compact among the nine elements that constitute the landscape of Rongga Sub-district.

The total edge (TE) is an absolute measure of the total edge length of a particular element type. Table 8 show that the highest TE value was obtained for the dry land forest element, with a total edge length of 738,075 m. The dry forest element was more elongated compared to the other elements, which is consistent with its high CA, PLAND, and LSI values. The lowest TE value in this study was obtained for the river element due to its small segment in the study area.

Based on the ED values, the elements making up the landscape of Rongga Sub-district, arranged from highest to lowest value, are as follows: secondary forest > open fields > agroforestry > shrubs > plantations > rainfed rice fields > settlements > irrigated rice fields > rivers. The highest ED value in the Rongga Sub-district landscape was that of the dry land forest element at 63.05 m/ha. This shows that the dry land forest had the most irregular edge among the elements in Rongga Sub-district.

3.2.4. Element Distributions

The interspersion and juxtaposition index (IJI) was used as a metric to determine the element distribution. The IJI measures the extent to which different patch types are in proximity to each other [57]. The results of the analysis using Fragstat 4.2.1 with the IJI metric are shown in Table 9.

Table 9. Calculation results of IJI metric at the class level.

No.	Type	IJI (%)
1	Agroforestry	75.92
2	Secondary Forest	77.83
3	Open Fields	85.68
4	Plantations	83.55
5	Settlements	90.51
6	Irrigated Rice Fields	90.62
7	Rainfed Rice Fields	89.35
8	Shrubs	72.41
9	Rivers	38.31

Abbreviation: IJI, interspersion and juxtaposition Index.

As can be seen from Table 9, the highest IJI value was for irrigated rice fields with a percentage of 90.62%, followed by settlements (90.51%) and rainfed rice fields (89.35%). IJI = 100% occurs when all types of patches are equally close to other patches. This suggests that irrigated rice fields were in close proximity to the other eight patches. In this study, the IJI percentage of the rivers was quite low (38.31%) compared to the other eight elements. The adjacency of the rivers, vis-à-vis the other elements, was uneven. Based on the classification of the land cover map, the river segments were only located in particular parts of the study area, such as in Sukaresmi and Bojong Salam villages.

3.3. Structural Analysis at the Landscape Level

At the landscape level, analyses were carried out to identify the distribution of elements and heterogeneity. The metrics used were NP, the Shannon Evenness Index (SHEI), and the Shannon Diversity Index (SHDI), and the results are shown in Table 10.

Table 10. Calculation results of NP, SHEI, and SHDI metrics at the landscape level.

Landscape	NP	SHEI	SHDI
Rongga Sub-district	2734	0.87	1.92

Abbreviations: NP, number of patches; SHEI, Shannon Evenness Index; SHDI, Shannon Diversity Index.

3.3.1. Element Distribution

At the landscape level, the Shannon Evenness Index (SHEI) measures 0 and 1. The SHEI equals 0 if the landscape consists of only one patch (i.e., there is no diversity) and approaches 0 if the area distribution of the patch types becomes uneven (i.e., dominated by only one kind of patch).

Based on the results obtained, the SHEI value in the landscape of Rongga Sub-district was 0.87, which is close to 1; therefore, it can be said that the distribution of patches in this landscape was quite even. This means that none of the landscape elements dominated the study area.

3.3.2. Heterogeneity

Heterogeneity is a measure of how the parts of a landscape differ from one another [58]. A high degree indicates that the distribution of various landscape types is almost even, and there is relatively high heterogeneity. The Shannon Diversity Index (SHDI) represents the heterogeneity of the landscapes in a study area [59].

Based on the results obtained, the SHDI value in Rongga Sub-district was 1.92, showing that the heterogeneity of this landscape was quite high. Habitat fragmentation occurring in a landscape can increase the number of land plots and the value of the SHDI [60]. This condition was supported by the NP value of 2734.

3.4. Ecological Networks at the Landscape Level

The implemented least-cost modeling resulted in a visual model of connectivity for multiple target species in the study area. The map of resistance surface distribution based on the experts' judgment and combined pathways that will benefit the multiple target species is shown in Figure 4. The most favorable area networks were identified in the center of the species' core areas. These networks link the eastern part and center of the region. We identified 11 ecological corridors connecting areas 1, 2, 3, 4, 5, 7, and 9 as the least-cost paths for the multiple target species' movement in the study area. The lowest cost surface was measured from these particular core areas in terms of the resistance value. The most favorable vegetation cover with diverse tree species mostly exists in the remnant forest of the extreme landscape. These vegetation structures and compositions facilitate the movement of the targeted species, especially the arboreal species such as *Nycticebus javanicus*, and connect two or more larger areas of wildlife habitats. Although the center has relatively steep slopes, the area might have greater value due to its dominant forest cover. This resulted in the relatively low resistance value of the landscape matrices in the area.

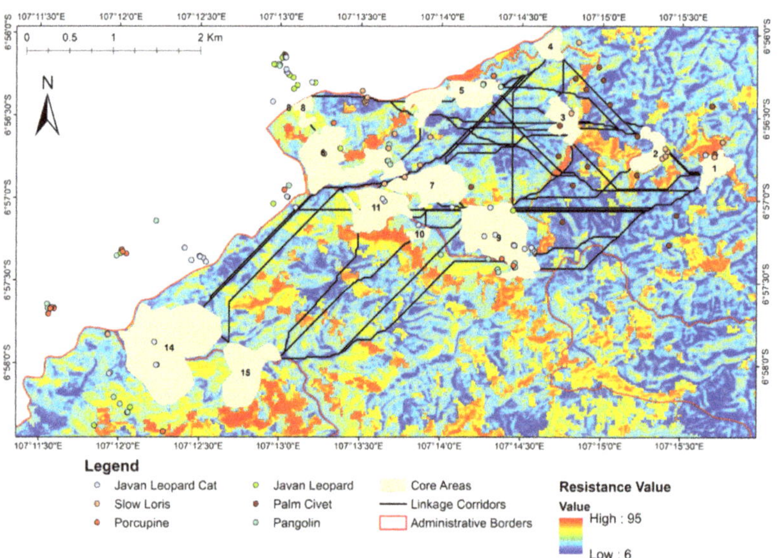

Figure 4. Map of resistance values and ecological network among core areas.

The dry land forest located in the western part of the study area is more fragmented than in the eastern region. This is due to the fact that the western part consists of various land use patches of different sizes, such as shrubs, bare land, plantations, settlements, and rainfed rice fields. The ecological networks in the western part connect areas 6, 7, 9, 11, 14, and 15. The directions of the least-cost paths generally follow the most low-cost surface area. In some cases, the existence of wildlife corridors that pass through land cover with high resistance values is inevitable—for example, the networks connecting areas 1, 2, and 3 to area 9. In order to access area 9, every target species has to pass through an access road. This is in line with the findings of a previous study which found that certain species such as *Nycticebus javanicus*, *Paradoxurus hermaphroditus*, and *Manis javanica* were found passing the access road [61].

Rongga's landscape, comprising agroforestry, secondary forests, and plantations, caused the contrast degree to be relatively low. This condition, together with the relatively good connectivity between the secondary forests, plantations, and agroforestry, is beneficial for the presence of wild animals needing heterogenous landscapes where the changes between one element and another are not drastic. It is clear that the extreme topography condition of Rongga Sub-district supports the landscape connectivity and, consequently, the presence of wild animals in this area (Figure 5). The relatively good connectivity might arise from the fact that the extreme landscape makes this area less accessible for humans and their activities.

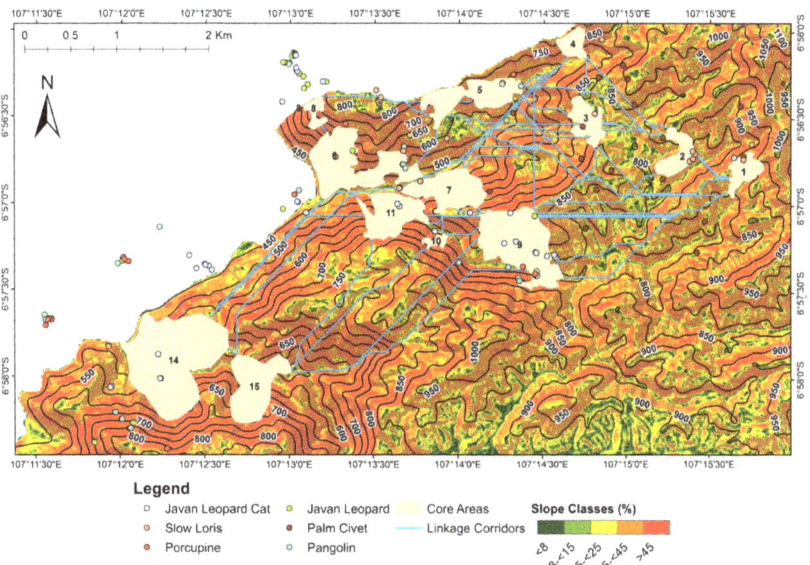

Figure 5. Species distribution and linkage corridors map related to slope gradient.

3.5. Ecological Function

Agroforestry is a collective name for land-use systems and technologies where woody perennials (trees, shrubs, palms, bamboo, etc.) are deliberately used on the same land management units as agricultural crops and/or animal habitats, in some form of spatial arrangement or temporal sequence [62]. Agroforestry landscapes are defined as multiple land-use systems or a combination of forestry and agricultural landscapes that are managed to create a balance between agricultural intensification and forest sustainability [63,64]. Agroforestry land-use systems have the potential to increase agricultural land use while providing lasting benefits and reducing adverse environmental impacts at the local and global levels. This system promotes increased productivity and environmental stability by reducing emissions from deforestation and forest degradation [65].

As a system that combines trees and/or shrubs (perennial) with agronomic crops (annual or perennial), agroforestry can sequester carbon both above and below ground. Such agroforestry systems play an important role in increasing carbon stocks in the terrestrial biosphere [66].

One study revealed the existence of common palm civets in talun (mixed) gardens, which are the most suitable habitat type [16]. Palm civet cats eating palm fruit were found in agroforestry/talun gardens. This suggests that agroforestry provides food for the common palm civets. Another study revealed the existence of Javan slow lorises in Rongga Sub-district [17]. The Javan slow loris was mostly found around talun vegetation: sengon talun and mixed talun. This proves that this type of land use has the potential to become this animal's habitat.

Secondary forests play an important role in conserving biodiversity, saving unique and endemic species that are adaptable to extreme conditions, and providing important ecosystem goods (e.g., livestock feed, firewood, medicine, and trade goods such as resin and sap) and services (e.g., formation and conservation of soil, conservation and quality improvement of water, setting of the water regime and microclimate, reducing the speed of the wind, control of wind erosion, and deceleration of the depletion of water) [67].

Figure 6 show some research that has revealed the presence of several animals in the dry land forests of Rongga Sub-district [16–20]. This research was specifically conducted in the future Cisokan sub-watershed area. One study successfully revealed the existence of the Javan leopard in both natural and production forests in Rongga Sub-district [18]. Furthermore, research revealed the existence of Javan leopards in natural forests [20]. The natural forests of Batu Nagok and Sarongge are far from human activities, and the main habitat of the Javan leopard is densely vegetated forests that are difficult to access, as well as areas with a steep topography (slope > 40%) and remote areas such as deep valleys or high hills that are difficult to reach.

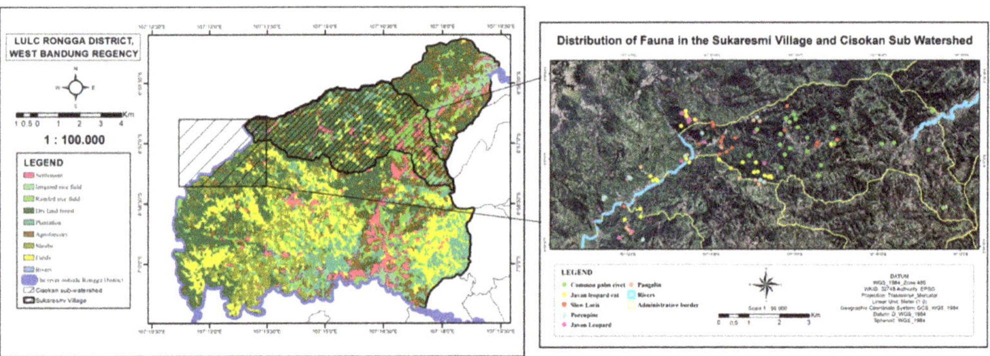

Figure 6. Land use and distribution of fauna in Rongga Sub-district.

One study revealed the existence of pangolins in Cisokan [19]. The points where pangolins were found in Rongga Sub-district were Batu Wulung, Curug (waterfall) Japarana, and Curug Walet. Additionally, subsequent research revealed the existence of slow lorises in Cisokan [17]. The point where slow lorises were found in Rongga Sub-district was in the Cilengkong area with secondary forest.

Dry land forests in Rongga Sub-district consist of both plantation and natural forests [40]. The plantation forests are pine (*Pinus merkusii*) production forests managed by Perhutani. Natural forests in Rongga Sub-district are mostly found on riverbanks along the Cisokan River, and the remaining natural forests are in Cigowek. Several types of plants that make up this natural forest, including *Ficus* sp., *Piper aduncum*, *Artocarpus elasticus*, *Macaranga tanarius*, and *Spatodea campanulata*, were found growing in river forests. Some of the typical forest tree types are *Dysoxylum parasiticum*, *Dipterocarpus hasseltii*, *Ficus retusa*, *Artacarpus elasticus*, *Ficus variegata*, and other *Ficus* sp.

Open field/moorland cover is used for dry land farming, especially for vegetables, chilies, and cassava [41]. The ecological function of farm fields is to mimic the structure of tropical forests, producing a lot of humus litter and burning biomass, which is very important as a nutrient for soil fertility. Farm fields also have a multilayered canopy that is structured stratigraphically and can withstand soil erosion [68]. At the landscape level, the cultivation system can keep the land well covered with vegetation, which helps reduce surface runoff and regulate water discharge [69].

A garden (plantation) is tree-growing land that is limited by ownership or other rights, has a canopy cover dominated by fruit or industrial trees, and has clear and regular boundaries [70]. Plantations in a landscape have important environmental benefits and play a role in sustainable production and improvement of soil quality, mitigation of water quality and carbon salinity, and biodiversity benefits [71]. In addition, plantations provide protection and a food source for local fauna and can even enhance the natural restoration of native forests [72,73]. Plantation patches can also improve landscape connectivity, acting as a species movement medium between remnants of the natural forest [74].

A settlement has a positive impact on the economic life of its residents but a negative effect on traditional village culture and the ecological landscape [75]. Human interaction with land through land uses such as building settlements and agriculture can affect the capacity of soil carbon storage, which is an important carbon reservoir that can be released as CO_2 into the atmosphere [76].

The emergence of new settlements causes water absorption systems to be disrupted, drainage networks to not function properly, and household waste from the surface to be accumulated [77]. Based on the results of the landscape structure analysis, the area of settlement in Rongga Sub-district is 6.393%, showing that the landscape of Rongga Sub-district has not experienced much human intervention; thus, the ecological function of other patches was not disturbed. A study revealed the existence of common palm civets in the residential area of Sukaresmi Village, Rongga Sub-district [16], and other research revealed the existence of many Javan slow lorises in the vicinity of Rongga Sub-district settlements [17]. This proves that this type of land use has the potential to become a habitat.

Rice fields have various ecological functions; for instance, they can replace natural wetlands with artificial wetlands so that they act as habitats for freshwater animals, breeding areas, shelters, feeding places, other services for wildlife [78], and as oxygen producers, contributing to the conservation of land and water [79]. The rice field ecosystem also functions as a conservation area that supports the hydrological process, as a flood mitigation area that provides retention reservoirs, as an area that creates a microclimate and reduces pollutants, as a public recreation area, and as a disaster mitigation/evacuation area [80].

The rice field structure, which has a flat surface and is flanked by embankments, makes rice fields function as small dams to collect rainwater, thereby reducing the possibility of flooding. The ability to hold rainwater was initially intended to provide sufficient water for rice plants at the growth stage [81]. Thus, rice fields are analogous to wetlands as temporary places for rainwater [80]. Therefore, a rice fields' ability to withstand rainwater and/or surface runoff is only useful when rain occurs to prevent flooding in lower areas or downstream [82].

Rice fields can also provide water purification. Rice fields can purify water if the incoming irrigation water contains high concentrations of nitrogen (N) and phosphorus (P). This purification occurs when the incoming nitrogen (N) concentration is 2–3 mg N/L or greater. The decrease in N concentration is caused by its reuse for crops and the denitrification of nitrate/nitrite-N in rice fields and irrigation/drainage systems [83].

Shrubs are used as ecological indicators because they form the majority of subcanopy structural layers in a forest. Shrub cover can indicate the habitat quality and a number of complex ecological processes that are interconnected [84]. Shrubs provide refuge and food for forest organisms. They also provide an input of essential organic materials to the ground, play a principal role in the nutrient cycle, contribute substantially to the diversity of compositions and structures, help protect watershed areas from erosion, and improve the aesthetics of a forest ecosystem [72]. For example, the coverage and distribution of shrubs affect the diversity and abundance of mycorrhizal fungi, which constitute important food for small mammals, which are important prey for avian and terrestrial predators [85].

Research has revealed the existence of common palm civets (*Paradoxurus hermaphroditus*) on the shrub cover in Sukaresmi Village in Rongga Sub-district [16]. This type of civet uses shrubs as a hiding place so as not to be seen by predators when looking for its food, especially at noon. Shrubs are the most suitable habitat type for civets due to the amount of food available for common palm civets in shrub-type habitats. Another study revealed that porcupines in the Cisokan sub-watershed were mostly found in shrubs [86].

Rivers have been widely used for settlements, infrastructure, and production. Rivers can provide drinking water, irrigation, fish as a food supply or for recreational fishing, and areas for flood protection [87]. Rivers play a role in regulating the flow of water and minerals originating from the surrounding land and influence the flow of materials and water [88]. Rivers carry soil and sediment from one place to another, which has a major impact on the landscape. The silt that settles in the river flood plains is channeled to several other elements, one of which is agricultural land, and agricultural land thus becomes fertile [89].

In addition, river flows play a role in the movement of geochemical and biological matter and energy in the environment and can become a habitat for river biota adapted to seasonally fluctuating flows. Rivers also provide spatial connectivity between habitats and allow for the spread of plants, animals, and fungi [90].

4. Conclusions

Based on our analysis, it can be concluded that the landscape of Rongga Sub-district consists of patch elements of agroforestry, open fields, settlements, shrubs, plantations, and rainfed as well as irrigated rice fields, with the dry land forest as a matrix and rivers as corridors. The prevalence of each element varies, with the highest score obtained for agroforestry, indicating a high degree of fragmentation. The patch shape was irregular and not compact, and the element distribution was fairly even: each patch was adjacent to other patch types, and the landscape heterogeneity was quite high. The vegetated landscape elements in the study area suggest that despite it being an extreme landscape, it might exhibit ecological functions, particularly in terms of wildlife protection and natural disaster prevention (e.g., against landslides and erosion).

Therefore, maintaining forest cover in the study area should be taken into account in natural resource management at the landscape level to safeguard the landscape integrity. Many elements of the Rongga Sub-district landscape play a role in environmental conservation, as indicated by the flow of energy and materials towards the patch level, as well as the mosaic level, and the presence of wildlife in some elements.

The use of Landsat images in this research means that the interpretation of the landscape was fragmented, which is a limitation of this study. The use of high-resolution orthophotos in future studies can help to gain a different interpretation of the cultural and complex landscapes.

Author Contributions: Conceptualization, S.W.; methodology, S.W., A.D.M. and M.A R.; software, A.D.M. and M.A.R.; validation, S.W., A.D.M. and M.A.R., P.P.; formal analysis, S.W., A.D.M. and M.A.R.; investigation, S.W., A.D.M., M.A.R.; resources, S.W., A.D.M. and M.A.R.; data curation, S.W., A.D.M. and M.A.R.; writing—original draft preparation, S.W; writing—review and editing, S.W., A.D.M. and P.P.; visualization, S.W., A.D.M. and M.A.R.; supervision, P.P.; project administration, S.W.; funding acquisition, P.P. All authors have read and agreed to the published version of the manuscript.

Funding: This research was funded by Ministry of Education and Culture through Universitas Padjadjaran.

Institutional Review Board Statement: Not applicable.

Informed Consent Statement: Not applicable.

Data Availability Statement: This study does not report any data.

Acknowledgments: The authors would like to thank the Indonesian Ministry of Education and Culture for supporting this research through *Penelitian Dasar, Penelitian Terapan Unggulan Perguruan Tinggi,* and Universitas Padjadjaran, Sumedang, Indonesia, for providing research funding through the Universitas Padjadjaran Academic Leadership Grant scheme.

Conflicts of Interest: The authors declare no conflict of interest.

References

1. Turner, M.G.; Gardner, R.H. *Landscape Ecology in Theory and Practice*; Springer: New York, NY, USA, 2015; Volume 401.
2. Wu, J.G. Landscape Ecology. In *Encyclopedia of Theoretical Ecology*; Hastings, A., Gross, L., Eds.; University of California Press: Berkeley, CA, USA, 2012; pp. 392–396.
3. Council of Europe. *European Landscape Convention, Florence 20. 10*; European Treaty Series: Cardiff City, UK, 2000; No. 176; p. 7.
4. Atik, M.; Işikli, R.C.; Ortaçeşme, V.; Yildirim, E. Definition of landscape character areas and types in Side region, Antalya-Turkey with regard to land use planning. *Land Use Policy* **2015**, *44*, 90–100. [CrossRef]
5. Ingegnoli, V. *Landscape Ecology: A Widening Foundation*; Springer: Singapore, 2002.
6. Tlapáková, L.; Stejskalová, D.; Karasek, P.; Podhrázská, J. Landscape Metrics as a Tool for Evaluation Landscape Structure—Case Study Hustopeče. *Eur. Countrys.* **2013**, *5*, 52. [CrossRef]
7. Mitchell, C.; Reich, P.; Tilman, D.; Groth, J.V. Effects of elevated CO_2, nitrogen deposition, and decreased species diversity on foliar fungal plant disease. *Glob. Chang. Biol.* **2003**, *9*, 438–451. [CrossRef]
8. Brodie, J.F.; Redford, K.H.; Doak, D.F. Ecological Function Analysis: Incorporating Species Roles into Conservation. *Trends Ecol. Evol.* **2018**, *33*, 840–850. [CrossRef]
9. van den Berg, E.; Santos, F.A.M. Aspectos da variação ambiental em uma floresta de galeria em Itutinga, MG, Brasil. *Ciência Florestal.* **2003**, *13*, 83–98. [CrossRef]
10. Budke, J.C.; Jarenkow, J.A.; De Oliveira-Filho, A.T. Intermediary disturbance increases tree diversity in riverine forest of southern Brazil. *Biodivers. Conserv.* **2010**, *19*, 2371–2387. [CrossRef]
11. Neary, D.G.; Ice, G.G.; Jackson, C.R. Linkages between forest soils and water quality and quantity. *For. Ecol. Manag.* **2009**, *258*, 2269–2281. [CrossRef]
12. Clark, D.B.; Clark, D.A.; Rich, P.M.; Weiss, S.; Oberbauer, S.F. Landscape-scale evaluation of understory light and canopy structures: Methods and application in a neotropical lowland rain forest. *Can. J. For. Res.* **1996**, *26*, 747–757. [CrossRef]
13. Forman, R.T.T. *Land Mosaics: The Ecology of Landscapes and Regions*; Cambridge University Press: Cambridge, UK, 1995.
14. Zhao, Q.; Wen, Z.; Chen, S.; Ding, S.; Zhang, M. Quantifying Land Use/Land Cover and Landscape Pattern Changes and Impacts on Ecosystem Services. *Int. J. Environ. Res. Public Health* **2019**, *17*, 126. [CrossRef] [PubMed]
15. Verma, A. Necessity of Ecological Balance for Widespread Biodiversity. *Indian J. Biol.* **2020**, *4*, 158–160. [CrossRef]
16. Parikesit; Withaningsih, S.; Prastiwi, W.D. Estimated Abundance and Distribution of Common Palm Civet (Paradoxurus hermaphroditus, Pallas 1777) in the Rural Landscape of Sukaresmi, West Bandung Regency. *IOP Conf. Ser. Earth Environ. Sci.* **2019**, *306*, 012003. [CrossRef]
17. Withaningsih, S.; Parikesit; Ayundari, A.; Prameswari, G.; Megantara, E.N.; Husodo, T. Distribution and habitat of Javan slow loris (Nycticebus javanicus É. Geoffroy, 1812) in non-conservation area. *AIP Conf. Proc.* **2018**, *2019*, 060006.
18. Shanida, S.S.; Partasasmita, R.; Husodo, T.; Parikesit, P.; Megantara, E.N. Short Communication: Javan Leopard Cat (*Prionailurus bengalensis javanensis* Desmarest, 1816) in the Cisokan non-conservation forest areas, Cianjur, West Java, Indonesia. *Biodiversitas* **2018**, *19*, 37–41. [CrossRef]
19. Withaningsih, S.; Noorahya, F.; Megantara, E.N.; Parikesit, P.; Husodo, T. Nest existences and population of Pangolin (Manis javanica Desmarest, 1822) at the designated area of Cisokan Hydropower, West Java, Indonesia. *Biodivers. J. Biol. Divers.* **2018**, *19*, 153–162. [CrossRef]
20. Shanida, S.S.; Partasasmita, R.; Husodo, T.; Parikesit, P.; Febrianto, P.; Dan Megantara, E.N. Short Communication: The existence of Javan Leopard (Panthera pardus melas Cuvier, 1809) in the non-conservation forest areas of Cisokan, Cianjur, West. Java, Indonesia. *Bodiversitas* **2018**, *19*, 42–46. [CrossRef]
21. West Bandung Regency Government. Medium-Term Investment Program Plan. West Bandung Regency, 2017–2022. 2015. Available online: https://sippa.ciptakarya.pu.go.id/sippa_online/ws_file/dokumen/rpi2jm/DOCRPIJM_cc06df8a09_BAB%20IBab%201%20RPIJM%20Kab%20Bandung.pdf (accessed on 18 November 2021).
22. West Bandung Regency Government. Medium-Term Investment Program Plan. West Bandung Regency, 2015–2019. 2008. Available online: https://sippa.ciptakarya.pu.go.id/sippa_online/ws_file/dokumen/rpi2jm/DOCRPIJM_3a3c8d3e3b_BAB%20IIBAB%202.pdf (accessed on 17 November 2021).

23. BPS. Rongga Subdistrict in Figures 2021. BPS-Statistics of Bandung Barat Regency. 2021. Available online: https://bandungbaratkab.bps.go.id/publication/2021/09/24/205c99047ff726dd984d59d0/kecamatan-rongga-dalam-angka-2021.html (accessed on 19 November 2021).

24. Disdukcapil. Geographic Information System (GIS) Kab.Bandung Barat Kecamatan Rongga and 8 Village. 2019. Available online: http://disdukcapil.bandungbaratkab.go.id/Home/gis/rongga (accessed on 15 August 2020).

25. Merchant, J.W.; Narumalani, S. Integrating Remote Sensing and Geographic Information Systems. In *The SAGE Handbook of Remote Sensing*, 1st ed.; Warner, A.T., Nellis, M.D., Foody, G.M., Eds.; SAGE Publications Ltd.: London, UK, 2009; pp. 257–268, ISBN 978-1-4129-3616-3.

26. Wischmeier, W.H.; Smith, D.D. *Predicting Rainfall Erosion losses—A Guide to Conservation Planning*; U.S. Department of Agriculture Science and Education Administration: Washington, DC, USA, 1978.

27. Prasetyo, L.B. *Pendekatan Ekologi Lanskap Untuk Konservasi Biodiversitas*; Faculty of Forestry, IPB University: Bogor, Indonesia, 2017.

28. Forman, R.; Gordon, M. *Landscape Ecology*; John Wiley and Son: New York, NY, USA, 1986.

29. Ministry of Forestry of Republic Indonesia. *Regulation of the Director General of Watershed Management and Social Forestry on Guidelines for Identification of Watershed Characteristics*; Ministry of Forestry: Jakarta, Indonesia, 2013.

30. Boehner, J.; Selige, T. Spatial Prediction of Soil Attributes Using Terrain Analysis and Climate Regionalisation. In *SAGA—Analysis and Modelling Applications*; Boehner, J., Ed.; Verlag Erich Goltze GmbH: Gottingen, Germany, 2006.

31. El Jazouli, A.; Barakat, A.; Ghafiri, A.; El Moutaki, S.; Ettaqy, A.; Khellouk, R. Soil erosion modeled with USLE, GIS, and remote sensing: A case study of Ikkour watershed in Middle Atlas (Morocco). *Geosci. Lett.* **2017**, *4*, 25. [CrossRef]

32. Yustika, R.D.; Somura, H.; Yuwono, S.B.; Arifin, B.; Ismono, H.; Masunaga, T. Assessment of soil erosion in social forest-dominated watersheds in Lampung, Indonesia. *Environ. Monit. Assess.* **2019**, *191*, 726. [CrossRef]

33. McGarigal, K.; Cushman, S.A.; Ene, E. FRAGSTATS v4: Spatial Pattern Analysis Program for Categorical and Continuous Maps. Computer Software Program Produced by the Authors at the University of Massachusetts, Amherst. 2012. Available online: http://www.umass.edu/landeco/research/fragstats/fragstats.html (accessed on 19 January 2020).

34. Gurrutxaga, M.; Lozano, P.J.; del Barrio, G. GIS-based approach for incorporating the connectivity of ecological networks into regional planning. *J. Nat. Conserv.* **2010**, *18*, 318–326. [CrossRef]

35. Etherington, T.R. Least-Cost Modelling and Landscape Ecology: Concepts, Applications, and Opportunities. *Curr. Landsc. Ecol. Rep.* **2016**, *1*, 40–53. [CrossRef]

36. Tang, Y.; Gao, C.; Wu, X. Urban Ecological Corridor Network Construction: An Integration of the Least Cost Path Model and the InVEST Model. *ISPRS Int. J. Geo-Inf.* **2020**, *9*, 33. [CrossRef]

37. Liu, C.; Newell, G.; White, M.; Bennett, A.F. Identifying wildlife corridors for the restoration of regional habitat connectivity: A multispecies approach and comparison of resistance surfaces. *PLoS ONE* **2018**, *13*, e0206071. [CrossRef]

38. Holt, C.; Watts, K.; Bellamy, C.C.; Nevin, O.; Ramsey, A. Defining Landscape Resistance Values in Least-Cost Connectivity Models for the Invasive Grey Squirrel: A Comparison of Approaches Using Expert-Opinion and Habitat Suitability Modelling. *PLoS ONE* **2014**, *9*, e112119. [CrossRef]

39. Walker, R.; Craighead, L. *Analysing Wildlife Movement Corridors in Montana Using GIS*; ESRI User Conference: San Diego, CA, USA, 1997.

40. Permana, S.; Iskandar, J.; Parikesit, P.; Husodo, T.; Megantara, E.N.; Partasasmita, R. Changes of ecological wisdom of Sundanese People on conservation of wild animals: A case study in Upper Cisokan Watershed, West Java, Indonesia. *Biodivers. J. Biol. Divers.* **2019**, *20*, 1284–1293. [CrossRef]

41. Yusuf, S.M.; Murtilaksonoc, K.; Hidayatc, Y.; Suharnotod, Y. Analysis and Prediction of Land Cover Changes in the Upper Citarum Basin. *J. Pengelolaan Sumberd. Alam Dan Lingkung.* **2018**, *8*, 365–375.

42. Tarolli, P.; Preti, F.; Romano, N. Terraced landscapes: From an old best practice to a potential hazard for soil degradation due to land abandonment. *Anthropocene* **2014**, *6*, 10–25. [CrossRef]

43. Pijl, A.; Tosoni, M.; Roder, G.; Sofia, G.; Tarolli, P. Design of Terrace Drainage Networks Using UAV-Based High-Resolution Topographic Data. *Water* **2019**, *11*, 814. [CrossRef]

44. Rudiarto, I.; Doppler, W. Impact of land use change in accelerating soil erosion in Indonesian upland area: A case of Dieng Plateau, Central Java—Indonesia. *Int. J. AgriSci.* **2013**, *3*, 558–576.

45. Chaplot, V.A.M.; Le Bissonnais, Y. Runoff Features for Interrill Erosion at Different Rainfall Intensities, Slope Lengths, and Gradients in an Agricultural Loessial Hillslope. *Soil Sci. Soc. Am. J.* **2003**, *67*, 844–851. [CrossRef]

46. Jordan, G.; van Rompaey, A.; Szilassi, P.; Csillag, G.; Mannaerts, C.; Woldai, T. Historical land use changes and their impact on sediment fluxes in the Balaton basin (Hungary). *Agric. Ecosyst. Environ.* **2005**, *108*, 119–133. [CrossRef]

47. Wandi, Y.; Manuwoto, S.; Buchori, D.; Hidayat, P.; Budiprasetyo, L. Spatial Analysis of Agricultural Landscapes and Diversity of Hymenoptera in Cianjur. *HAYATI J. Biosci.* **2006**, *13*, 137–144.

48. Bargali, S.S. Forest Ecosystem: Structure and Functioning. *Curr. Trends For. Res.* **2018**, *2*, 1–3. [CrossRef]

49. Ministry of Forestry of Republic Indonesia. *Strategic Plan. 2010–2014*; Ministry of Forestry: Jakarta, Indonesia, 2010.

50. Mäler, K.G.; Aniyar, S.; Jansson, A. Accounting for Ecosystem Services as a Way to Under Stand the Requirements for Sustainable Development. *Proc. Natl. Acad. Sci. USA* **2008**, *105*, 9501–9506. [CrossRef] [PubMed]

51. Goodwin, B.J.; Fahrig, L. How does landscape structure influence landscape connectivity? *Oikos* **2002**, *99*, 552–570. [CrossRef]

52. Pozo-Montuy, G.; Serio-Silva, J.C.; Bonilla-Sanchez, Y.M. The influence of the matrix on the survival of arboreal primates in fragmented landscapes. *Primates* **2011**, *52*, 139–147. [CrossRef]
53. Martins, R.N.; Abrahão, S.A.; Ribeiro, D.P.; Colares, A.P.F.; Zanella, M.A. Spatio-Temporal Analysis of Landscape Patterns in the Catolé Watershed, Northern Minas Gerais. *Rev. Árvore* **2018**, *42*, 1–11. [CrossRef]
54. Ji, J.; Wang, S.; Zhou, Y.; Liu, W.; Wang, L. Spatiotemporal Change and Landscape Pattern Variation of Eco-Environmental Quality in Jing-Jin-Ji Urban Agglomeration from 2001 to 2015. *IEEE Access* **2020**, *8*, 125534–125548. [CrossRef]
55. McGarigal, K. Fragstats Help. 2015. Available online: https://www.umass.edu/landeco/research/fragstats/documents/fragstats.help.4.2.pdf (accessed on 19 January 2020).
56. Fynn, I.E.; Campbell, M.J. Forest Fragmentation Analysis from Multiple Imaging Formats. *J. Landsc. Ecol.* **2019**, *12*, 1–15. [CrossRef]
57. Saritha, S.; Kumar, G.S. Analysis of the smart growth of kochi city through landscape metrics. In Proceedings of the 2017 IEEE Region 10 Symposium (TENSYMP), Cochin, India, 14–16 July 2017; pp. 1–5. [CrossRef]
58. Gunawan, H.; Prasetyo, L.B. *Forest Fragmentation: The Teory Underlying Forest Spatial Planning towards Sustainable Development*; Research and Development Center for Conservation and Rehabilitation: Bogor, Indonesia, 2013.
59. Lv, J.; Ma, T.; Dong, Z.; Yao, Y.; Yuan, Z. Temporal and Spatial Analyses of the Landscape Pattern of Wuhan City Based on Remote Sensing Images. *ISPRS Int. J. Geo-Inf.* **2018**, *7*, 340. [CrossRef]
60. Herzog, F.; Lausch, A. Supplementing land-use statistics with landscape metrics: Some methodological considerations. *Environ. Monit. Assess.* **2001**, *72*, 37–50. [CrossRef] [PubMed]
61. PLN. Biodiversity Management Plan (BMP) Monitoring Species Upper Cisokan Pumped Storage. UCPS, Bandung, Indonesia. 2017, *unpublished work*.
62. Nair, P.R. *An Introduction to Agroforestry*; Kluwer Academic Publisher: Dordrecht, The Netherlands, 1993.
63. Kaswanto, F.; Tataq, A.; Choliq, M.; Bagus, S. Revitalizing Rural Landscape Yards as Landscape Service Providers to Improve Community welfare. *J. Lanskap Indones.* **2016**, *8*, 50–60. [CrossRef]
64. Owonubi, J.J.; Otegbeye, G.O. Disappearing forest: A review of the Challenges for Conservation of genetic resources and environmental management. *J. For. Res. Manag.* **2002**, *1*, 1–11.
65. Sobola, O.O.; Amadi, D.C.; Jamala, G.Y. The Role of Agroforestry in Environmental Sustainability. *J. Agric. Vet.* **2015**, *8*, 20–25. [CrossRef]
66. Jacob, D.E.; Ufot, I.N.; Sotande, A.O. Climate Change Adaptation and Mitigation through Agroforestry Principles in the Sahal Region of Nigeria. In Proceedings of the 35th Annual Conference of the Forestry Association of Nigeria, Sokoto, Nigeria, 11–16 February 2013; pp. 300–308.
67. Forest Conservation Team. *Guidelines on Sustainable Forest Management in Drylands of Sub-Saharan Africa. Arid Zone Forests and Forestry Working Paper*; No. 1; FAO: Rome, Italy, 2010.
68. Iskandar, J. *Cultivation Ecology in Indonesia*; Djambatan: Jakarta, Indonesia, 1989.
69. Mulyoutami, E.; van Noordwijk, M.; Sakuntaladewi, N.; dan Agus, F. *Changes in Cultivation Pattern: A Shifting Perception of Cultivators in Indonesia*; World Agroforestry Centre—ICRAF, SEA Regional Office: Bogor, Indonesia, 2010; p. 101.
70. Badan Standarisasi Nasional. Klasifikasi Penutup Lahan SNI No 7645:2014. Bagian 1: Skala Kecil dan Menengah. 2014. Available online: www.bsn.go.id (accessed on 18 April 2020).
71. Australian Goverment. Plantations and Farm Forestry. 2019. Available online: https://www.agriculture.gov.au/forestry/australias-forests/plantation-farm-forestry (accessed on 17 November 2020).
72. Tellería, J.L.; Galarza, A. Avifauna Invernante En Un Eucaliptal Del Norte De España. *Ardeola* **1991**, *38*, 239–247.
73. Humphrey, J. Benefits to biodiversity from developing old-growth conditions in British upland spruce plantations: A review and recommendations. *Forestry* **2005**, *78*, 33–53. [CrossRef]
74. Brockerhoff, E.G.; Jactel, H.; Parrotta, J.A.; Quine, C.P.; Sayer, J. Plantation forests and biodiversity: Oxymoron or opportunity? *Biodivers. Conserv.* **2008**, *17*, 925–951. [CrossRef]
75. Wang, Z.; Wang, C.; Jiang, Z.; Hu, T.; Han, W.; Zhang, C.; Jin, J.; Wei, K.; Zhao, J.; Wang, X. Relationship between Rural Settlements' Plant Communities and Environmental Factors in Hilly Area of Southeast China. *Sustainability* **2020**, *12*, 2771. [CrossRef]
76. Liang, M. Interaction of Human Settlement, Vegetation Patterns and Soils in the Northern Minnesota Prairie-Forest Region from the Euro-American Settlement Era to Present. Ph.D. Thesis, University of Wisconsin-Madison, Madison, WI, USA, 2017.
77. Mukhoriyah. Study of the Ecological-Economic Value of Rice Fields and Its Relation to Spatial Planning in Depok City. Master's Thesis, Universitas Indonesia, Jakarta, Indonesia, 2012.
78. Zong, Y.; Chen, Z.; Innes, J.B.; Chen, C.; Wang, Z.; Wang, H. Fire and flood management of coastal swamp enabled first rice paddy cultivation in east China. *Nat. Cell Biol.* **2007**, *449*, 459–462. [CrossRef]
79. Santosa, I.; Gusti, N.; Adnyana, G.M.; dan Dinata, I.; Ketut, K. The Impact of Rice Field Land Function Change on Rice Food Security. In *Prosiding Seminar Nasional Budidaya Pertanian | Urgensi dan Strategi Pengendalian Alih Fungsi Lahan Pertanian | Bengkulu 7 July 2011*; Universitas Bengkulu: Bengkulu, Indonesia, 2011; ISBN 978-602-19247-0-9.
80. Natuhara, Y. Ecosystem services by paddy fields as substitutes of natural wetlands in Japan. *Ecol. Eng.* **2013**, *56*, 97–106. [CrossRef]
81. Yoshikawa, N.; Nagao, N.; Misawa, S. Evaluation of the flood mitigation effect of a Paddy Field Dam project. *Agric. Water Manag.* **2010**, *97*, 259–270. [CrossRef]

82. Masumoto, T.; Yoshida, T.; Kubota, T. An index for evaluating the flood-prevention function of paddies. *Paddy Water Environ.* **2006**, *4*, 205–210. [CrossRef]

83. Maruyama, T.; Hashimoto, I.; Murashima, K.; Takimoto, H. Evaluation of N and P mass balance in paddy rice culture along Kahokugata Lake, Japan, to assess potential lake pollution. *Paddy Water Environ.* **2008**, *6*, 355–362. [CrossRef]

84. Muir, P.S.; Mattingly, R.L.; Tappeiner, J.C., II; Bailey, J.D.; Elliot, W.E.; Hagar, J.C.; Miller, J.C.; Peterson, E.B.; Starkey, E.E. *Managing for Biodiversity in Young Douglasfir Forests of Western Oregon. Biological Science Report USGS/I3RD/BSR-2002-0006*; US Geological Survey, Biological Resources Division: Corvallis, OR, USA, 2002.

85. Carey, A.B.; Kershner, J.; Biswell, B.; Dominguez De Toledo, L. Ecological Scale and Forest Development, Squirrels, Dietary Fungi, And Vascular Plants in Managed and Unmanaged Forest. *Wildl. Monogr.* **1999**, *63*, 3–71.

86. Mustikasari, I.A.; Withaningsih, S.; Megantara, E.N.; Husodo, T.; Parikesit, P. Population and distribution of Sunda porcupine (*Hystrix javanica* F. Cuvier, 1823) in designated area of Cisokan Hydropower, West Java, Indonesia. *Biodivers. J. Biol. Divers.* **2019**, *20*, 762–769. [CrossRef]

87. Böck, K.; Polt, R.; Schülting, L. Ecosystem Services in River Landscapes. In *Riverine Ecosystem Management*; Schmutz, S., Sendzimir, J., Eds.; Springer: Cham, Switzerland, 2018; Volume 8, pp. 413–433. [CrossRef]

88. Wuisang, C.E.V.; dan Dwight, M.R. Greenbelt Planning in Urban Riverbank Landscapes. In *Prosiding Temu Ilmiah IPB*; Institut Pertanian Bogor: Bogor, Indonesia, 2015; pp. 103–108.

89. Wuisang, C. Biodiversity Conservation in Urban Areas: Landscape Evaluation of Green Corridors in Manado City. *Media Matrasain* **2015**, *12*, 47–60.

90. Mims, M.C.; Olden, J. Life history theory predicts fish assemblage response to hydrologic regimes. *Ecology* **2012**, *93*, 35–45. [CrossRef]

Article

An Indicator-Based Approach to Assess the Readiness of Urban Forests for Future Challenges: Case Study of a Mediterranean Compact City

Mᵃ Fernanda Maradiaga-Marín [1] and Paloma Cariñanos [1,2,*]

[1] Department of Botany, University of Granada, 18071 Granada, Spain; mfmaradiaga@correo.ugr.es
[2] Andalusian Institute for Earth System Research (IISTA-CEAMA), University of Granada, 18071 Granada, Spain
*   Correspondence: palomacg@ugr.es

Abstract: Urban Forests (UFs) are key elements in Mediterranean compact cities, as they provide numerous ecosystem benefits and increase the resilience of cities against the anticipated impacts of climate change. It is, thus, necessary to review all the aspects that may have a negative effect on their ecosystem functions and the services that they provide. In this paper, a set of indicators is proposed that allow for a preliminary evaluation of some of the main disservices and factors that Mediterranean UFs present and the ways to maximize their benefits for users. For this purpose, 20 indicators, divided into three categories—Biodiversity, Accessibility/Facilities, and Infrastructure—were selected. Within these three categories, a range of values was established, from low to high or absence/presence. The indicators were tested in 24 urban forests of different types, all of which are representative of a medium-sized compact Mediterranean city. The results highlight that the UFs have adequate species richness and diversity, but among the species present are quite a few that emit BVOCs and allergens, as well as some that have invasive behavior. Poor cleaning, absence of night lightning, and scarcity of water points are aspects to improve in a good number of UFs; while a high surface area of impermeable soil, low tree cover, and extensive areas of grass that require large amounts of water for maintenance are the main issues of the infrastructure block that need medium-term planning to be addressed. It can be concluded that the proposed set of indicators allows for a general assessment of the readiness of UFTs in Mediterranean cities for the upcoming climatic, social, and ecological challenges.

Keywords: ecosystem disservices; Mediterranean region; green infrastructure; urban forest types; indicators

Citation: Maradiaga-Marín, M.F.; Cariñanos, P. An Indicator-Based Approach to Assess the Readiness of Urban Forests for Future Challenges: Case Study of a Mediterranean Compact City. *Forests* **2021**, *12*, 1320. https://doi.org/10.3390/f12101320

Academic Editors: Manuel Esperon-Rodriguez and Tina Harrison

Received: 18 August 2021
Accepted: 24 September 2021
Published: 27 September 2021

Publisher's Note: MDPI stays neutral with regard to jurisdictional claims in published maps and institutional affiliations.

1. Introduction

An urban forest (UF) can be described as all the trees and wooded areas in an urban space. This includes both public and private spaces, such as parks and gardens, and alongside streets, homes, and schools. Overall, UFs improve the urban ecosystem and increase recreational value, so they can be considered as key elements in the interconnected network that forms urban green infrastructure (UGI) [1,2]. As integral components of urban ecosystems, UFs provide a variety of Ecosystem Services (ES) [3–5]. Moreover, they bolster urban connectivity [6] and social cohesion and integration [7], provide shelter in situations of risk and natural disasters [8], and have been fundamental elements of physical and mental health restoration during the COVID-19 pandemic [9]. Given the significant impact that climate change can have in some areas of the world, UFs also stand out as one of the most efficient measures for strengthening climate resilience in cities [10]. Hence, increasing and reinforcing urban greening is one of the main solutions being implemented in municipalities to mitigate and adapt to the effects of climate change [11]. However, this process of greening cities may be slowed down by the urban, socio-cultural, and structural limitations of each territory. Thus, in densely populated cities it may be difficult to find areas to develop new green spaces. In this way, already existing UFs must be prepared for the current and future environmental, social, economic, and cultural challenges, so that they can guarantee the population access to high-quality spaces that improve life quality.

The Mediterranean region is currently one of the most urbanized areas in the world and has some of the fastest urbanization rates. With growth rates of most major cities between 1.5% and 2% per year [12], the current percentage of the population that live in cities (55%) is expected to increase to 68% by 2050, testing the capacity of Mediterranean cities to manage growth [13,14]. Additionally noteworthy is the lack of a sustainable urban development model to anticipate rapid population growth, which is causing a densified and extended urbanization towards the periphery. However, above all, this growth is causing a change in the traditional characteristic compact growth [15]. A problem related to this high urban densification is the lack of available space for the implementation of green areas [16]. The use of any available surface as buildable land or the extension of the city to the (sometimes scarce) peripheral natural areas is causing not only losses in terms of available green surface, but also making it difficult for certain social sectors to access UFs [17]. Furthermore, the scarcity of green areas is characteristic of many Mediterranean compact cities, since many of these cities have low per capita green space [18]. This situation highlights that, given the difficulty of finding areas for the implementation of new UFs, the planning and management of existing ones must be prioritized to maximize the ecosystem functions they are providing.

It is common for many Mediterranean cities to have prioritized cultural and tourism infrastructure when designing and planning new urban green spaces [19]. The Mediterranean climate, which is characterized by mild temperatures for many months of the year and an amount of sunshine greater than 2500 h/year, favors many outdoor activities [20], with UFs being the center of positive social, intercultural, and intergenerational interactions that cultivate social cohesion in ways that enhance health and well-being [7]. The rich historical and cultural heritage of Euro-Mediterranean cities can also be seen in the high number of Historic Gardens and other Green Spaces of different Mediterranean countries in the UNESCO World Heritage Cultural Landscapes program [21]. However, there are other relevant aspects that have not received the same attention, and that in one way or another affect both the quality of these spaces and the benefits to the population, as well as a positive net balance of ecosystem services. For various reasons (aesthetic, traditional) or due to lack of planning, UFs have been the entry point for pests and diseases that threaten biodiversity [22]. They can be the origin for the expansion of invasive species [23] or the source of the emission of compounds (BVOCs and allergens) with an impact on health [24,25]. Frequently, the bioclimatic characteristics of the Mediterranean climate have not been considered when designing thermal comfort spaces [26] or selecting plant species with water requirements compatible with local species [27]. On other occasions, the scarcity of equipment and services, the absence of facilities for vulnerable groups, or impediments to accessibility to the space are some of the factors that affect user satisfaction [28]. Thus, a review of all these aspects that can negatively impact the balance of ecosystem services provided by UFs in Mediterranean cities is necessary.

Monitoring the ecosystem services provided by urban forests through indicators is one of the best ways to quantify and value the capacity of natural processes and components to provide goods and services that satisfy human needs [29]. Some authors have highlighted how this provisioning capacity is variable according to a series of structural, morphological, location, scale, and management factors. These factors are related both to urban forests themselves and to the urban environment [30,31]. The incorporation of indicators to measure ecosystem services as an approach for risk assessment or on the impact on human well-being is becoming more frequent [30,32,33]. In these proposals, researchers usually use either a single category that groups the different disservices [30] or different categories are established depending on the type of disservice or hazard they cause: Aesthetic, Safety, Health, Economic, and Environmental, among others [34,35]. The impacts caused by anthropization in green areas have also been evaluated, and two categories of indicators have been established: conservation indicators and public use indicators [36]. In addition to the indicators to assess ecosystem services, it is also important to add indicators to assess other factors like what the conditions are for the inhabitants who visit these spaces for recreation. The consideration of the benefits that these green spaces can provide would not be complete without considering certain amenities such as lighting, recreational elements, and bathroom facilities, among others [28,37].

Understanding the ecological context and the factors that influence the use and enjoyment of UFs is related to socio-cultural characteristics and needs. Thus, this work aimed to establish a series of indicators of some main disservices and factors that UFs of Mediterranean cities can present, so that they can be used as a preliminary checklist for fine-tuning and improvement needs, and to reinforce the functions and ecosystem services they provide. This work is justified by the significant impact that climate change will have on the Mediterranean region—declared one of the most vulnerable to the effects of increased temperatures and reduced rainfall—as well as the significant changes in city models that are expected in the coming decades, where the shortage of new green areas will continue to increase.

2. Materials and Methods

2.1. Study Area

The study was carried out in the city of Granada (37.179937, −3.603489; 680 m.a.s.l), located in the southeast of the Iberian Peninsula (Figure 1), which presents the characteristics of the typical model of a medium-sized Mediterranean compact city. These characteristics include a high population density, compact buildings, complex urban uses, the proximity of services on a pedestrian scale, and mixed residential and commercial areas [38,39]. Granada contains a population density of 2654.4 inhabitants/km^2 (233,648 inhabitants, 88.02 km^2 of surface) and a 21.8% of the population is above 65 years of age [40,41]. Agriculture, commerce, and scientific–technical activities are the main sectors of economic activity [42]. The city is in a wide depression formed by the Genil River and the Sierra Nevada Mountain Range valley and has a continental-Mediterranean climate with an average annual temperature of 15.6 °C and an annual rainfall of 359 mm (for the period 1981–2010). The city also has a wide range of thermal amplitude [43]. Granada has been identified as one of the Mediterranean cities showing greater frequency of heat and cold waves [44]. Due to the intensity of traffic in and around the city, and the numerous sources of bioaerosol-type pollen emissions, Granada is currently one of the most polluted cities in Spain by both biotic and abiotic pollutants [45,46].

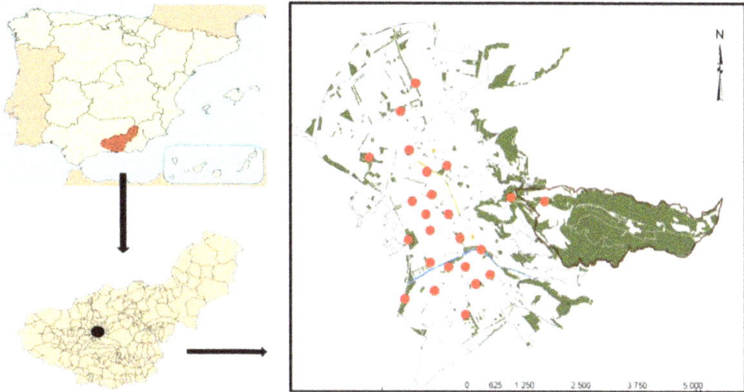

Figure 1. Location of the city of Granada along with the location of the 24 urban green spaces considered in this study [39].

Selection of Cases

According to data provided by the Environmental Information Network of Andalusia (REDIAM), there are 363 registered UFs of different types in the city, although 341 of them (93.9%) have a surface area less than 10,000 m^2 [38]. The total green surface area is 1,141,884.7 m^2, which translates to a public green surface area of 4.9 m^2 per inhabitant. For our study, 24 of these UFs with a total area of 543,583 m^2 (47.6% of the total green area of the city) were considered. The criteria for selecting the urban forests included those that are in public spaces and are mostly managed by the local Administration or, failing that, by other public administrations, so that all aspects related to its management and maintenance were regulated by the same regime. We were also interested in selecting UFs located within residential and commercial neighborhoods, frequently visited by the population and/r tourists, due to its historical character, and at the same time can potentially support urban biodiversity and provide recreational services. Finally, we also considered as a criterion that the UFs were representative of compact/historical Mediterranean cities; that is, places in which the typologies city parks, pocket parks, tree-lined streets, boulevards, and plazas are the most abundant.

2.2. Selection of Indicators

To check for possible disservices and other factors that UFTs could present, we selected a series of simple measurement indicators—those that can be used by researchers and laypeople alike, which can also be replicated in any green space (either ones that are similar or different to those in our study). A range of values were applied to each indicator, which were established according to some existing scales and reference literature [32,34,47]. Each of these parameters and the range of values to be applied are detailed below. For better functionality, the indicators were grouped into three categories: Biodiversity, Accessibility/Facilities, and Infrastructure.

2.2.1. Biodiversity Indicators

This category reviews the indicators related to the species' biodiversity of green areas. This includes an indicator that measures the abundance of species and an indicator that refers to the negative effects that these species may have, such as allergenicity, toxicity, invasiveness, or emission of BVOCs. A summary of the indicators that make up this block is presented in the following table (Table 1):

Table 1. Values to assign to the indicators of the biodiversity category.

%	Species Richness [1]	Shannon' Index [2]	Allergenic Species [3]	Toxic/Poisonous Species [4]	Invasive Species [5]	Non- Native Species [6]	BVOCs Emission [7]
Low	<2	<2	<20	<10	<1	<10	<20
Moderate	2–10	2–3	20–30	10–15	1–2	10–50	20–30
High	>10	>3	>30	>15	>2	>50	>30

1: Species richness: This represents the number of different species that exist in each UFT, considering each space as an ecological community. It does not consider the abundance of the species. When considering UFs of different types, among which there may be promenades or boulevards, often monospecific, we assigned a value of <2 for the lower range and a value of >10 to express a high richness [48].

2: Shannon Index: The Shannon Diversity Index is a quantitative measure that allows for the species' biodiversity (the relative abundance of each species) in a community to be estimated. In natural ecosystems, its value varies between 0.5 and 5, although an acceptable value is between 2 and 3; values that are less than 2 are considered low in diversity and values that are greater than 3 are high in species diversity [49]. We applied this rank to the UFTs considered since they are components of an Urban Ecosystem.

3: Allergenic species: To set the ranges for this category, those species that have a moderate or high Allergenic Potential Value (APV) according to the allergenicity scale proposed by Cariñanos et al., 2014, 2017, and Cariñanos and Marinangeli, 2021 [50–52] were considered. A percentage higher than 20% of allergenic species in a given UF is enough to be considered an allergenic risk for the sensitive population.

4: Toxic/Poisonous species: UFs are spaces visited by all kinds of social groups, some of them very vulnerable, such as children. The fact that 10% of plant species in UFs have toxic substances may already represent a moderate risk to this population. When 15% of species in UFs have toxic substances, it is considered to be a maximum risk [53,54].

5: Invasive species: The incorporation of exotic species as ornamentals in UFs has become one of the main entry ways for species that, after escaping from urban areas, have settled in the surrounding areas and thereby behave as invasive species [55]. For this indicator, we considered that a single species has sufficient potential to be considered a moderate risk [56].

6: Non-native species: Non-native species are considered those that have been introduced by humans [57]. Due to the long history that the incorporation of non-native species has had in Mediterranean urban forests [58], and the constancy of the invasive behavior that some of these species have developed [55], we considered that the categories established for this variable should be considered a potential disservice. When less than 10% of the total species in a given UF is non-native, the value is low. When the number is above 10%, the value is considered to be moderate.

7: BVOCs' emissions: BVOCs' emissions have been classified according to the percentage of existing species with low (<1), moderate (<10), or high emissions (>10) of BVOCS in micrograms VOC g Dry Weight $-1\,h^{-1}$, according to existing reference scales [59,60]. The values assigned to the different categories of this service are similar to those of allergenic species: <20% for low values and >30% for high values.

2.2.2. Accessibility/Facilities Indicators

In this category, factors that can positively affect both user comfort and satisfaction are considered, so the indicator only expresses the presence (YES) or absence (NO) of these factors. Doing this highlights the existence of an aspect in need of improvement. If a specific indicator is marked as absent, this would have a negative effect for a particular population group such as the disabled, the elderly, children, or women (for example, accessibility or the existence of sports equipment or playgrounds), or on all population sectors (for example, the absence of any type of water element that provides a practical, aesthetic, or recreational functionality). The indicators that were considered for this section are listed below (Table 2):

Table 2. Proposed indicators for checking the Accessibility/Facilities category.

	YES	NO
Accessibility [8]		
Water points [9]		
Playgrounds [10]		
Night lighting [11]		
Outdoor fitness equipment [12]		
Sport facilities [13]		
Rest area s[14]		
Cleaning [15]		

8: Accessibility: Universal access to UFs has been considered a right for all citizens. This indicator considers the existence of physical barriers that hinder access to vulnerable social groups: the disabled, the elderly, or people with reduced mobility in general [61].

9: Water points: Water points in UFs can have different functionalities, although in this category the practical (drinking fountains), aesthetic, or recreational services that they can offer to visitors were considered. Therefore, a total absence of water points should be subject to improvement in the short term. [62].

10: Playgrounds: Due to the proven health benefits that the presence of playgrounds in UFs have for children [63], the results of this indicator will allow for the planning of playground implementation in the most appropriate spaces according to the characteristics of the UF itself and the socio-economic environment in which it is located.

11: Night lighting: The scarcity or absence of night lighting has been labelled as a very negative factor due to the fear of crime and the insecurity it produces in UFs [64]. Its absence is notorious in those UFs that can be accessed at night.

12: Outdoor fitness equipment: The existence of outdoor fitness equipment encourages the performance of physical activity for the aged population. Thus, the absence of fitness equipment can negatively affect this population sector [65].

13: Sport facilities: The availability of sport facilities in UFs is one of the recreational services most appreciated by visitors due to the positive impact that sports can have on physical and mental health. As a result, the absence of sports facilities, even for things such as walking or running, can be a negative factor when visiting an UF [66].

14: Rest areas: One of the activities most valued by UF visitors is the contemplation of nature, relaxation, rest, and reflection. It is therefore necessary that rest areas, in the form of benches or other types of equipment, be available. [36]. Together with the accessibility indicator, the availability of rest areas is one of the most important factors to be addressed in case that they are absent.

15: Maintenance/Cleaning: The new maintenance strategies of UFs indicate that cleaning/dirt avoidance actions are one of the most successful improvement measures towards sustainable maintenance [67] and user satisfaction [66].

2.2.3. Gray-Green-Blue Infrastructures' Indicators

In this category, indicators were established for all aforementioned characteristics of UFs that, either due to the structural arrangement of their components or their superficial extension, can impede a correct provision of ecosystem services. They contemplate the three spheres of infrastructures that may be present in UF: blue, green, and gray. For the quantification of each indicator, i-Tree software was used. The i-Tree is a tool developed by the USDA Forest Service, and it is made up of different utilities that allow for the evaluation and quantification of the benefits and values that certain trees produce in urban environments [68]. The i-Tree Canopy tool is designed to estimate the percentage of vegetation cover and other cover classes within each study area. This tool places random points on Google Earth satellite images and determines the coverage class that the random point falls into. For our purposes, the following coverage categories were established: Tree/shrub cover and Grass/herbaceous surface (green infrastructure); Buildings and other gray infrastructure, Impervious soil (gray infrastructure); and Water elements in the form of small lakes, ponds, or ditches (which is different from the water points of the previous category) (blue infrastructure). The ranges to be applied to each indicator are presented in Table 3.

Table 3. Indicators proposed for checking the infrastructure category.

%	Canopy/Shadow Provision [16]	Impervious Soil [17]	Blue ELEMENTS [18]	Grass Surface [19]	Gray Infrastructure [20]
Low	>60	<20	>10	<5	<10
Moderate	40–60	20–40	1–10	5–10	10–15
High	<40	>40	<1	>10	>15

16: Canopy/Shadow provision: To establish the range of values of this parameter, the percentage of continuous surface and the spatial configuration of the trees were considered. Due to the high temperatures that can be registered during the summer in Mediterranean cities, a value of 40% of surface covered with trees was established as a minimum value for the effect of UFs on thermal comfort to be effective [69,70].

17: Impervious soil: The high surface area of impervious soil in cities has been a cause of flooding and a decrease in water quality. The scale of values for this indicator was established on the basis that a waterproof coverage of 10% may increase up to 20% due to surface runoff [71].

18: Blue elements: The existence of blue infrastructure in UFs has an effect on mitigating the urban heat island effect, but also on mitigating pollution by both biological and non-biological particulate matter. Its total absence or low surface coverage can, therefore, be a relevant disservice [72,73].

19: Grass surface: The area covered with grass in areas with Mediterranean climate requires a large amount of water for maintenance, in addition to making it difficult for other species to form natural grasslands. A narrow range of values were established for the quantification of the service: an area greater than 10% of the total is sufficient for the applicable value to be the maximum [74].

20: Gray infrastructure: In a general way, gray infrastructure is comprised of constructed assets that occupy land. Then, the larger the area covered by gray infrastructure in a park, the less effect it will have on the provision of ecosystem services [75].

3. Results

In this work, indicators were used to check the characteristics that may affect the provision of ecosystem services and user comfort in 24 UFTs in the city of Granada. Of the 24 selected UFs, 11 of them have a surface area of less than 10,000 m^2, which mainly includes typologies such as squares, tree alignments, boulevards, and pocket parks, and which are among the best represented in the city of Granada. The remaining UFs have an area that ranges between 13,500 (Quinta Alegre Garden) and 71,500 m^2 (García Lorca Park). This category includes some gardens, city parks, and the Bosque de Gomerez, one of the urban forests that surrounds the monument of La Alhambra. Regarding their management, 20 are managed by the Park and Garden Service of the Granada City Council, two are managed by the University of Granada (the Arboretum and the Botanical Garden of the Faculty of Law), and two are managed by the Board of the Alhambra (the Forest of Gomerez and the Carmen de los Martires). Other characteristics of UFs including species richness, number of trees, and percentage of allergenic trees are presented in Supplementary Material Table S1. The results obtained after applying the three categories of established indicators, Biodiversity, Accessibility / Facilities, and Infrastructure, are presented below.

3.1. Biodiversity Indicators

The application of the values assigned to each of the indicators of the Biodiversity category (Table 4) reveals that for indicator 1, most of the UFTs have an adequate species richness, with more than 10 different species in each, regardless of their typology. However, this is not transferable to the Shannon Diversity Index (indicator 2) since some parks have a value lower than 3, which can be considered as low diversity. In relation to the percentage of allergen-emitting species (indicator 3), it is notable that almost all types are above 30%, although this cannot be interpreted as a high Allergenicity Index value for the UFTs as a whole. Regarding the percentage of toxic species (indicator 4), the presence in some spaces of several species with some toxic part, such as seeds (*Malus* spp., *Prunus* spp.), bark, leaves, flowers, fruits (*Melia azedarach, Ligustrum lucidum, Robinia pseudoacacia, Aesculus hippocastanum*), or the whole plant (*Schinus molle, Taxus baccata*) is noteworthy. The indicator on the number of invasive species (indicator 5) considers the presence of only two species as a high value. Although most UFTs do not contain any, or only one, the total list of species of all the parks includes some of those with the greatest invasive potential: *Acacia dealbata, Ailanthus altissima, Eleagnus angustifolia, Eucalyptus camaldulensis, Gleditsia triacanthos*, and *Robinia pseudoacacia*. Closely related to this indicator is the percentage of non-native species (indicator 6), which is higher than 50% of all the species present in most of the green areas considered. The percentage of species emitting BVOCs (indicator 7) is also very high, which exceeds 30% of all species, with the most frequent being *Platanus* x *hispanica*, and several species of the *Populus* and *Quercus* genera, all of them highly represented in the different types of UFs. Similar to the allergenicity indicator, moderate- to high-emission species were considered.

Table 4. Results obtained for the Biodiversity indicators of the different UFs.

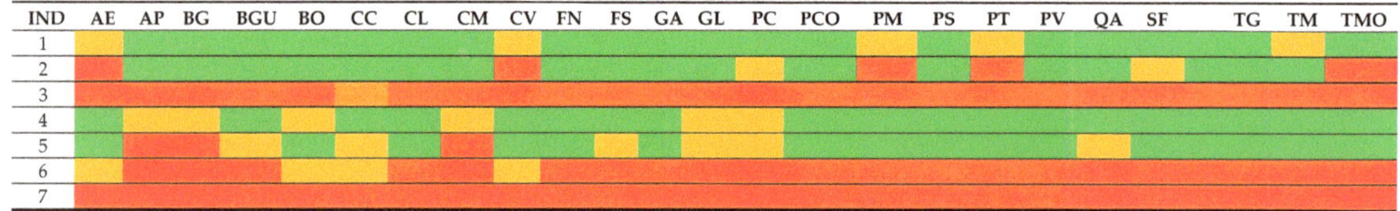

AE: Albert Einstein Plaza; AP: Almunia Park; BG: Gomerez Forest; BGU: University Botanical Garden; BO: Bola de Oro Park; CC: Carlos Cano Park; CL: Cruz de Lagos Park; CM: Carmen de los Martires Garden; CV: Carrera de la Virgen Park; FN: Fuentenueva Park; FS: Faculty of Sciences Arboretum; GA: Garcia Arrabal Park; GL: Garcia Lorca Park; PC: Concordia Plaza; PCO: Palace of Congresses Park; PM: Mariana Pineda Park; PS: Paseo del Salon Blvd.; PT: Trinidad Plaza; PV: Paseo del Violon Blvd.; QA: Quinta Alegre Garden; SF: San Fernando Plaza; TG: Triunfo Gardens; TC: Tico Medina Park; TMO: Tete Montoliu Park. Low ▮ Moderate ▮ High ▮.

3.2. Accessibility/Facilities Indicators

The factors checked in this category (Table 5) revealed that all UFTs have rest areas in the form of benches and rest places (indicator 14). Most UFTs (19 out of 24) are also accessible to population groups with reduced mobility, in which there are no obstacles to access or to have a walk inside (indicator 8). However, the absence of ramps and the presence of stairs in several UFs has a limiting effect. Poor cleaning (indicator 15) and the absence of night lightning (indicator 11) are important limitations in eight UFs, while the absence of water elements (indicator 9), which includes drinking fountains, is shared by 10 spaces. The major deficiencies in the UFTs are the absence of playgrounds (indicator 10) or outdoor fitness facilities (indicator 12), which are absent in 16 and 18 UFs, respectively. These deficiencies may cause displeasure to both children and the elderly.

Table 5. Results obtained for the Accessibility/Facilities indicators of the different UFs.

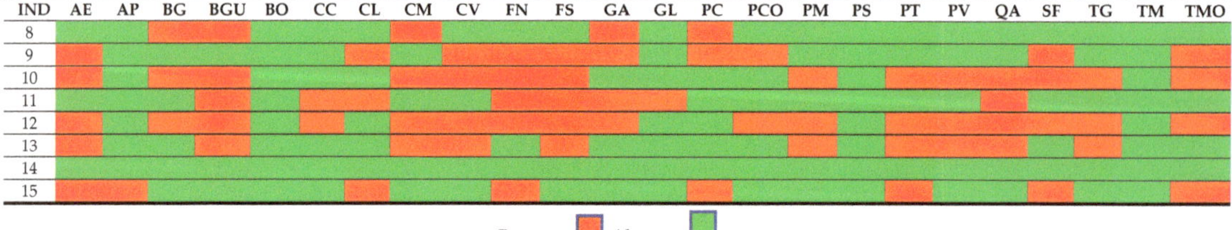

Presence ▮ Absence ▮.

3.3. Infrastructure Indicators

After applying the indicators related to infrastructure (Table 6), it was observed that water elements (indicator 18) are absent in 18 of 24 parks. The percentage of tree cover, that is, the continuous area covered by tree canopy (indicator 16), registers values of <40% in a high number of spaces. However, spaces such as the Gomerez Forest, University Botanical Garden, Carmen de los Martires Garden, Carrera de la Virgen Park, Mariana Pineda Park, and Trinidad Plaza have a canopy percentage of >60% of their total area. The percentage of impervious cover (indicator 17) is higher than 20% in 14 of out 24 UFs, but it reaches more than 40% in areas such as Albert Einstein Plaza. Although the majority of UFs have a grass coverage of less than 5% (indicator 19), in some, such as Gomerez Forest, Bola de Oro Park, Fuentenueva Park, Garcia Arrabal Park, Garcia Lorca Park, Triunfo Gardens, and Tico Medina Park, there are lawn areas greater than 10% of the total area. The existence of gray infrastructure elements is quite low, and it is only relevant in the University Botanical Garden, due to its location attached to the Faculty of Law and is moderate in the park surrounding the Palace of Congresses Park, where this building represents most of the gray infrastructure present.

Table 6. Results obtained for the Infrastructure indicators of the different UFs.

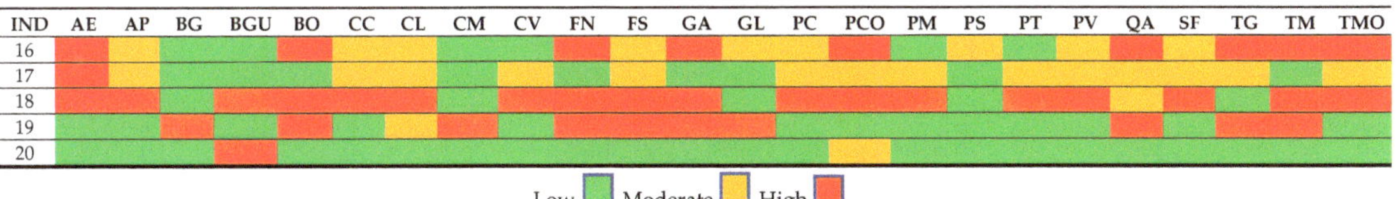

4. Discussion

In this work, we proposed indicators that allow for a preliminary check on the state of some of the functions and services provided by the UFTs of a medium-sized Mediterranean city. For this, a series of measurable values were assigned to 20 parameters of different categories, so that they can be applied in a simple way and be easily interpreted by experts and non-experts [76]. This proposal is inspired by other existing ones in which both ecosystem services and disservices were quantified through models, direct measurements, or from the reviewed scientific literature [29,33,77]. In this case, we adjusted the indicators to the particularities and idiosyncrasies of Mediterranean cities, characterized by their compactness, population density, and vulnerability to the effects of climate change [78]. The list of indicators of the existing services in UFTs aims to be an easy-to-use tool that highlights the improvement actions necessary to reinforce the role of green areas in the framework of a sustainable and resilient city [79].

For the assessment of disservices, 24 urban forests of different types were selected, which are the most characteristic of the Mediterranean region in which historical urban development has generated compact model cities with high population densities and little green area [15]. In this framework, spaces with a surface area greater than 10,000 m^2 are particularly relevant, represented in our list by City Parks, which have been identified as one of the urban forest types with the greatest significance to address some of the most important challenges at local and global level [1,80]. However, if there is one type of urban forest that should stand out for its eminent Mediterranean character, it is the small plaza types and squares (also called pocket parks). These spaces have a surface area of less than 1 hectare, are accessible, safe, and attractive, and allow people to carry out a large number of socialization and recreation activities [20,81]. Their abundance in numerous cities (for example, more than 90% of existing public green areas in Granada [38]), and their location in historic city centers [81,82] mean that a special focus should be made to review and fine-tune all those aspects that are affecting the provision of ecosystem services. The important demographic challenges that the Mediterranean region must face in the coming decades, with intra- and inter-country migratory flows and the region's vulnerability to climate change [83], should accelerate the planning of operational measures that make UFTs healthier, more visible, inclusive, cohesive, aesthetic, equitable, enjoyable, and, ultimately, sustainable places.

The 20 indicators were grouped into three categories to plan improvement joint interventions. In the category of Biodiversity indicators, it is evident that the emitting capacity of allergens and BVOCs of different plant species was not considered when selecting the species that form urban forests. Both services need urgent mitigation measures given their impact on health and the environment [84,85]. Some authors point out the diversification of the flora, the selection of individuals/strains with low pollen production, proper management, and knowledge of the functional traits of the species as solutions for reducing the impact of emissions of allergens and BVOCs [34,84–86]. The percentage of toxic and/or poisonous species in UFs does not represent a serious problem, since the most vulnerable groups have been adequately warned [87].

Several species used in UGI have invasive behavior and their use should be discontinued. They should be identified and controlled in nurseries to avoid their incorporation into UGI elements, and a specific and proactive handling of the specimens that are already installed in UFs are some of the actions to be carried out to avoid the spread and incorporation of new invasives. [24]. Priority attention should be paid to species such as *Ailanthus altissima*, already mentioned as one of the most invasive species in the Mediterranean region, and whose colonization in vulnerable ecosystems is considered an important threat to native species [88].

The Accessibility/Facilities block includes indicators that can improve the well-being and satisfaction of UFT users, although there are important differences between the spaces considered in this study. Thus, while in most of these spaces there are rest areas, mainly seats, and they are accessible to people with motor disabilities, playgrounds and fitness facilities are absent in a good number of them. These Park facilities have been shown to have a multifunctional role as elements for socialization and health within the framework of sustainable cities [89], so their incorporation into the design should be considered a priority in all those UFTs that have a primary social function, which would include those of the City Parks, Boulevards, and pocket parks located both close to historic districts and in new neighborhoods to serve a wide range of social groups. The installation of water elements is also urgent, specifically in the form of drinking fountains due to the impact that climate change may have in the Mediterranean region, where a significant increase in temperature and greater frequency and intensity of heat waves are expected [90]. The amount and area of the blue points (water) identified in most of the green spaces are not significant in ecological terms, since these points are not usually large enough to influence the local temperature of the studied areas. These water points have aesthetic purposes or come in the form of drinking fountains and, therefore, it is not considered that they could ameliorate urban heat island events. It is also important to review the current water services so that they can be used safely in situations such as those generated by the COVID-19 pandemic [91].

Two indicators that are included in this block deserve particular attention: poor cleaning and the absence of night lightning. Cleaning or proper management of public spaces is one of the most valued services when surveys of the user population are carried out [92,93]. Therefore, the presence of dirt derived from poor cleaning is a clear negative factor. In our study, it was observed that those UFTs furthest away from the Historic Center are those with the most garbage and plastic bottles on the ground. Full bins, graffiti, and animal feces are also common in these areas. The improvement of this service implies a greater investment of human and economic resources on the part of the Municipal Parks and Gardens Maintenance Services, which, in most cases, it does not have the capacity or flexibility to modify the traditional handling and maintenance actions that have been carried out for a long time [94]. However, perhaps now is the time to plan new management perspectives in which not only aesthetic and cultural services predominate but also align as closely as possible to the natural dynamics of these urban ecosystem components [95].

Security, specifically the lack of security in parks that are poorly lit at night, is one of the issues most frequently pointed out by the surveyed population. Walking or crossing isolated parks, at night and with poor lighting, are some of the main attributes that evoke fear of crime in public urban parks. Individual factors such as gender and past experiences compound this problem [96,97]. In our study, adequate night lighting is absent in some of the UFTs that usually close access for night visitors, such as the Faculty of Sciences, the University Botanical Garden, or the Quinta Alegre Garden, but also others that are in very populated districts such as the Cruz de Lagos Park, the Carlos Cano Park, or the Fuentenueva Park, in which these lighting characteristics should be reviewed.

The infrastructure indicators contemplate the presence and extension of the elements corresponding to the green, blue, and gray spheres. One of the most outstanding disservices is that of the blue elements in the form of small lakes, ponds, or monumental fountains. Although the presence of water elements has been a historical tradition closely linked to Mediterranean gardens, brought to their maximum splendor with the Italian Renaissance water gardens [98], their presence in current UFs has been greatly reduced. One of the reasons for this reduction is the importance of the water economy in the management of green areas, which is being imposed throughout the region as a result of the growing water deficit. This requires the implementation of new strategies for efficient water management, both to irrigate the spaces and to retain rainwater [99].

Closely related to this indicator are indicators 17 and 19, related to the percentage of impervious soil and surface area covered with grass. The amount of impermeable soil in UFTs is the origin of several important disservices such as the increase in runoff during periods of intense rainfall [71] or the exacerbation of the urban heat island effect [100]. Given the climate scenario forecasts for the Mediterranean region, with estimated temperature increases of up to 2 °C and a reduction of 30% in average annual precipitation [101], a transformation of these impervious surfaces into permeable soil is not only necessary but also urgent. On occasions, substituting compacted pavements with permeable pavements could lead to economic savings, with a significant return on investment when balancing costs due to flooding and replacement costs [102].

Regarding grass-covered areas, the progressive substitution of grass lawns by covers of native species with less water requirements is a debate already open in cities with Mediterranean climate [74,103]. Among the main Nature-based Solutions for the sustainable management of urban parks, the naturalization of the current lawns and grasslands is being contemplated, allowing for the introduction of native species that are better adapted to the prevailing climatic and ecological conditions [74,104]. The provision of ecosystem services, such as the conservation and improvement of biodiversity, the attraction of pollinators, the improvement of soil microorganisms, and the reinforcement of the population dynamics of the vegetation, would significantly mitigate the impact of the current disservice.

Finally, the results obtained for indicator 16, which establishes the percentage of continuous surface covered by tree canopy, were analyzed. This indicator can have different interpretations, since it is a value that is used for the assessment of ecosystem services, such as the mitigation of air pollution [105] or the reduction of the kinetic impact of raindrops [106]. In our case, since the purpose was to carry out a check on the current state of UFs and its immediate effect on the comfort of users, we associated the canopy percentage to the provision of shade, since one of the characteristics that can most prevent visits to UFs is the lack of thermal comfort. Our results have revealed that more than half of the UFs in Granada have a tree canopy of less than 40%, which is why they would also have a deficit in terms of the provision of shade. In the Mediterranean region, where the number of hours of sunshine/year is higher than 2500 [107] and UV radiation is one of the highest on the planet [108], the shaded spaces under tree canopies become authentic oases of shelter for a good number of hours a day [109]. These shaded spaces have also been shown to have a restorative effect on the mental health of the population, and to be cohesive and socializing elements in neighborhoods [110]. An improvement measure would entail the transformation of the extensive area of impervious soil that many of the parks have into a useful area for tree planting. This action would, in turn, take part in the greening process that many cities are undertaking as an urban resilience reinforcement measure, to address the impacts of climate change more adequately.

5. Conclusions

In this work, a series of indicators are proposed that allow for the status of some functions and services provided by urban forests of different typology to be checked. This work is extremely important since it allows for the evaluation of different future alternatives for better adaptation to climatic scenarios, urban growth, and population needs, which stakeholders can use in their decisions. Knowing the status of UFTs is an indicator of sustainable cities, where the management of public areas and ecosystem services, together with good environmental policies and development plans, are essential in areas affected by climate change, as is the case of Mediterranean cities.

Compact cities, such as Mediterranean urban areas with high population densities and a scarcity of green spaces, may have better urbanization plans in which, if the increase in urban green areas is not possible, there is the possibility of improving the quality of the green areas that already exist. Addressing the deficiencies of these areas based on the 20 proposed indicators may result in an improvement in aspects like the thermal comfort for the population or a reduction in cases of allergenicity.

It is also important to note that the selection of native arboreal species is important for maintenance costs. In those green spaces with a typology of city and pocket parks whose purpose is to provide recreation, it is important to visualize the aspects of improving accessibility and infrastructure facilities so that they can be used by all segments of the population. Likewise, it is not only the lack of blue infrastructure but considering their size and location within green areas would improve temperature and humidity.

The high percentage of impermeable soil is an aspect to be highlighted, since although it is necessary in some areas for accessibility, impervious areas were also found without any purpose, which could be replaced with vegetation, which allows for the filtration of rainwater. This is related to the fact that more than half of the UFTs studied have a percentage of less than 40% tree canopy. This lack of tree canopy is directly related to problems in mitigating urban heat islands and capturing polluted air.

Adopting strategies that are focused on future climate scenarios in Mediterranean cities is a key element in adapting public spaces to mitigate their own negative impacts on human health and ecosystem services. Therefore, the proposed set of indicators allows for a general assessment of the readiness of UFTs in Mediterranean cities to face the important climatic, social, and ecological challenges in the near future.

Supplementary Materials: The following are available online at https://www.mdpi.com/article/10.3390/f12101320/s1. Table S1: Characteristics of the 24 urban forests considered in this study.

Author Contributions: M.F.M.-M. and P.C. are both responsible for the conceptualization, methodology, formal analysis, investigation, resources, data curation, writing—original draft preparation, and writing—review and editing. All authors have read and agreed to the published version of the manuscript.

Funding: This research received no external funding.

Institutional Review Board Statement: Not Applicable.

Informed Consent Statement: Not Applicable.

Data Availability Statement: Characteristics of the 24 urban forests considered in this study.

Conflicts of Interest: The authors declare no conflict of interest.

References

1. Salbitano, F.; Borelli, S.; Conigliaro, M.; Chen, Y. *Guidelines on Urban and Periurban Forestry*; FAO Forestry Papers N; Food and Agriculture Organization of the United Nations: Rome, Italy, 2016.
2. Benedict, M. *How Cities Use Parks for Green Infrastructure*; City Parks Forum Briefing Papers; American Planning Association: Chicago, IL, USA, 2004.
3. Mexia, T.; Vieira, J.; Príncipe, A.; Anjos, A.; Silva, P.; Lopes, N.; Freitas, C.; Santos-Reis, M.; Correia, O.; Branquinho, C.; et al. Ecosystem Services: Urban parks under a magnifying glass. *Environ. Res.* **2018**, *160*, 469–478. [CrossRef]
4. Palliwoda, J.; Banzhaf, E.; Priers, J.A. How do the green components of urban green infrastructure influence the use of ecosystem services? Examples from Leipzig, Germany. *Landsc. Ecol.* **2020**, *35*, 1127–1142. [CrossRef]
5. Mao, Q.; Wang, L.; Guo, Q.; Li, Y.; Lin, M.; Xu, G. Evaluating Cultural Ecosystem Services of Urban Residential Green Spaces from the perspectives of Residents' Satisfaction with green spaces. *Front. Public Health* **2020**, *8*, 226. [CrossRef]
6. Torabi, N.; Lindsay, J.; Smith, J.; Khor, L.A.; Sainsbury, O. Widening the lens: Understanding urban parks as a network. *Cities* **2020**, *98*, 102527. [CrossRef]

7.   Jennings, V.; Bamkole, O. The relationship between social cohesion and urban green space: An Avenue for Health promotion. *Int. J. Environ. Res. Public Health* **2019**, *16*, 452. [CrossRef]

8.   Cariñanos, P.; Calaza, P.; Hiemstra, J.; Peralmutter, D.; Vilhar, U. The role of urban and peri-urban forests in reducing risks and mananging disasters. *Unasylva* **2018**, *69*, 53–58.

9.   Ugolini, F.; Masseti, L.; Calaza-Martínez, P.; Cariñanos, P.; Dobbs, C.; Krajter-Ostoic, S.; Marin, A.M.; Pearlmutter, D.; Saaroni, H.; Sauliene, I.; et al. Effects of the COVId-19 pandemic on the use and perceptions of urban Green space: An international exploratory study. *Urban For. Urban Green.* **2020**, *56*, 126888. [CrossRef] [PubMed]

10.   Corkery, L.; Marshall, N. Urban Parks and open space: Underpinning a city's future resilience. Analysis and Policy Observatory. In Proceedings of the 8th State of Australian Cities National Conference, Adelaide, Australia, 28–30 November 2018.

11.   Canals Ventín, P.; Lázaro Marín, L. *Towards Nature-Based Solutions in the Mediterranean*; IUCN Centre for Mediterranean Cooperation: Málaga, Spain, 2019.

12.   Woertz, E. Notes Internacionals CIDOB: Mediterranean Cities, the Urban Dimension of the European Neighborhood Policy. In *Mediterranean Cities, the Urban Dimension of the European Neighborhood Policy (6)*; CIDOB: Barcelona, Spain, 2018.

13.   Chaline, C. *Urbanisation and Town Management in the Mediterranean Countries. Assessment and Perspectives for Sustainable Urban Development*; Mediterranean Commission on Sustainable Development:UNEP-MAP/BP-RAC: Sophia Antipolis, France, 2001.

14.   Blue Plan. *Cities and Sustainable Development in the Mediterranean*; Mediterranean Comission on Sustainable Development: Sophia Antipolis, France, 2000.

15.   Salvati, L.; Morelli, V.G. Unveiling urban sprawl in the Mediterranean Region: Towards a latent Urban Transformation. *Int. J. Urban Reg. Res.* **2014**, *38*, 1935–1953. [CrossRef]

16.   Fullerm, R.A.; Gaston, K.J. The scaling of green space coverage in European cities. *Biol. Lett.* **2009**, *5*, 352–355. [CrossRef] [PubMed]

17.   Haaland, C.; van den Bosh, C.K. Challenges and strategies for urban green-spaces planning in cities undergoing densification: A Review. *Urban For. Urban Green.* **2015**, *14*, 760–771. [CrossRef]

18.   Kabish, N.; Haase, D. Green spaces of European cities revisited for 1990–2006. *Landsc. Urban Plan.* **2013**, *110*, 113–122. [CrossRef]

19.   Gospodini, A.; Brebbia, C.A.; Tiezzi, E. (Eds.) *The Sustainable City V: Urban Regeneration and Sustainability*; Wit Press: Southampton, UK, 2008; Volume 5.

20.   Adinolfi, C.; Suárez-Cáceres, G.P.; Cariñanos, P. Relation between visitors'behaviour and characteristics of green spaces in the city of Granada, south-eastern Spain. *Urban For. Urban Green.* **2014**, *13*, 534–542. [CrossRef]

21.   European Route of Historic Gardens. Cultural Routes of the Council of Europe. Available online: http://www.europeanhistoricgardens.eu (accessed on 18 August 2021).

22.   Tomlison, I.; Potter, C.; Bayliss, H. Managing tree pests and diseases in urban settings: The case of oak Processionary moth in London, 2006–2012. *Urban For. Urban Green.* **2015**, *14*, 286–292. [CrossRef]

23.   Gaertner, M.; Wilson, J.R.U.; Cadotte, M.W.; MacIvor, J.S.; Zenni, R.D.; Richardson, D.M. Non-native species in urban environments: Pattern, processes, impacts and challenges. *Biol. Invasions* **2017**, *19*, 3461–3469. [CrossRef]

24.   Calfapietra, C.; Fares, S.; Manes, F.; Morani, A.; Sgrigna, G.; Loreto, F. Role of Biogenic Volatile Organic Compounds (BVOC) emitted by urban trees on ozone concentration in cities: A review. *Environ. Pollut.* **2013**, *183*, 71–80. [CrossRef] [PubMed]

25.   Cariñanos, P.; Adinolfi, C.; de la Guardia, C.D.; de Linares, C.; Casares-Porcel, M. Characterization of allergen emission sources in urban areas. *J. Environ. Qual.* **2016**, *45*, 244–252. [CrossRef] [PubMed]

26.   Aram, F.; Solgi, E.; Baghaee, S.; Higueras-García, E.; Mosavi, A.; Band, S.S. How parks provide thermal comfort perceptions in the metropolitan cores; a case-study in Madrid, Mediterranean climate zone. *Clim. Risk Manag.* **2020**, *30*, 100245. [CrossRef]

27.   Heywood, V.H. The nature and composition of urban plants diversity in the Mediterranean. *Flora Mediterr.* **2017**, *27*, 195–220.

28.   Liu, R.; Xiao, J. Factors affecting users' satisfaction with urban parks through online comments data: Evidence from Shenzen, China. *Int. J. Environ. Res. Public Health* **2021**, *18*, 252.

29.   De Groot, R.; Wilson, M.A.; Boumans, R.M. A Typology for the classification, description and valuation of ecosystems functions. goods and services. *Ecol. Econ.* **2002**, *41*, 393–408. [CrossRef]

30.   Dobbs, C.; Escobedo, F.J.; Zipperer, W.C. A framework for developing urban forests ecosystem services and good indicators. *Landsc. Urban Plan.* **2004**, *99*, 196–206. [CrossRef]

31.   Davies, H.; Doick, K.; Handley, P.; O'Brien, L.; Wilson, L. Delivery of ecosystem services by urban forests. *Res. Rep. For. Comm. UK* **2017**, *26*, 77.

32.   von Döhren, P.; Haase, D. Ecosystem disservices research: A review of the state of the art with a focus on cities. *Ecol. Indic.* **2015**, *52*, 490–497. [CrossRef]

33.   von Döhren, P.; Haase, D. Risk assessment concerning urban ecosystem disservices: The example of street trees in Berlin, Germany. *Ecosyst. Serv.* **2019**, *40*, 101031. [CrossRef]

34.   Lyytimäki, J. Disservices of urban trees. In *Routledge Handbook of Urban Forestry*; Ferrini, F., van den Bosch, C.K., Fini, A., Eds.; Routledge Taylor and Francis: London, UK; New York, NY, USA, 2017.

35.   Cariñanos, P.; Calaza, P.; O'Brien, L.; Calfapietra, C. The cost of greening: Disservices of urban trees. In *The Urban Forests: Cultivating Green Infrastructure for People and the Environment*; Pearlmutter, D., Calfapietra, C., Samson, R., O'Brien, L., Krajter-Ostoic, S., Sanesi, G., Alonso, R., Eds.; Springer International Publishing AG: Cham, Switzerland, 2017.

36. Vieira Martins, L.F.; Bittas Venturi, L.A.; Belem Wingter, G. A monitoring system proposal for urban parks in Valley bottoms. *Ambiente Soc.* **2019**, *22*, e00243. [CrossRef]

37. Blaszczyk, M.; Suchocka, M.; Wojnowska-Heciak, M.; Myszynska, M. Quality of urban parks in the perception of city residents with mobility difficulties. *PeerJ* **2020**, *8*, e10570. [CrossRef]

38. Delgado-Capel, M.; Cariñanos, P. Towards a Standard Framework to Identify Green Infrastructure Key Elements in Dense Mediterranean Cities. *Forests* **2020**, *11*, 1246. [CrossRef]

39. IUCN. *Nature Based Solutions in Mediterranean Cities. Rapid Assessment Report and Compilation of Urban Nventories (2017–2018)*; IUCN: Málaga, Spain, 2019; p. 117.

40. UrbiStat. Maps, Analysis and Statistics about the Resident Population. Municipality of Granada. Available online: https://ugeo.urbistat.com/AdminStat/en/es/demografia/dati-sintesi/granada/20234622/4 (accessed on 12 September 2021).

41. National Institute of Statistics (INE). Población Residente por Fecha, Sexo y Edad. Resultados por Provincias. Available online: https://www.ine.es/jaxiT3/Tabla.htm?t=31304 (accessed on 31 July 2020).

42. Granada Instituto de Estadística y Cartografía de Andalucía; Consejeria de Transformación Económica, Industria, Conocimiento y Universidades; Junta de Andalucía. *Sistema de Información Multiterriorial de Andalucia (SIMA)*; Granada Instituto de Estadística y Cartografía de Andalucía: Granada, Spain; Consejeria de Transformación Económica, Industria, Conocimiento y Universidades: Seville, Spain; Junta de Andalucía: La Roda de Andalucía, Spain, 2021.

43. State Agency of Meteorology (AEMET). *Status Report of the Climate of Spain Executive Summary*; Ministry for the Ecological Transinction and Demographic Challenge, Government of Spain: Madrid, Spain, 2020.

44. Cuadrat, J.; Serrano-Notivoli, R.; Tejedor, E. Heat and Cold Waves in Spain. In *Fenómenos Meteorológicos Adversos en España*; Ediciones García Legaz, A.M.V., Rodríguez, V., Eds.; World Climate Research Programme: Geneva, Switzerland, 2013.

45. Casquero-Vera, J.A.; Titos, G.; Alados-Arboledas, L. *Diagnóstico de la Calidad del Aire del Área Metropolitana de Granada*; Agenda Ayuntamiento de Granada; Instituto Interuniversitario de Investigación del Sistema Tierra en Andalucía, Universidad de Granada: Granada, Spain, 2016.

46. Cariñanos, P.; Foyo-Moreno, I.; Alados, I.; Guerrero-Rascado, J.L.; Ruiz-Peñuela, S.; Titos, G.; Cazorla, A.; Alados-Arboledas, L.; Díaz de la Guardia, C. Bioaerosols in urban environments: Trends and Interactions with pollutants and meteorological variables based on quasi-climatological series. *J. Environ. Manag.* **2021**, *282*, 111963. [CrossRef]

47. Wu, S.; Li, B.V.; Li, S. Classifying ecosystem disservices and valuating their effects—A case study of Beijing, China. *Ecol. Indic.* **2021**, *129*, 107977. [CrossRef]

48. Gotelli, N.J.; Colwell, R.K. Estimating Species Richness. In *Biological Diversity: Frontiers in Measurement and Assessment*; Oxford University Press: Oxford, UK, 2011; Volume 39, p. 54.

49. Shannon, C.E.; Weaver, W. *The Mathematical Theory of Communication*; University Illinois Press: Urbana, IL, USA, 1949.

50. Cariñanos, P.; Casares-Porcel, M.; Quesada-Rubio, J.M. Estimating the allergenic potential of urban green spaces: A case-study in Granada, Spain. *Landsc. Urban Plan.* **2014**, *123*, 134–144. [CrossRef]

51. Cariñanos, P.; Casares-Porcel, M.; Díaz de la Guardia, C.; Aira, M.J.; Belmonte, J.; Boi, M.; Elvira-Rendueles, B.; De Linares, C.; Fernández-Rodriguez, S.; Maya-Manzano, J.M.; et al. Assessing allergenicity in urban parks: A nature-based solution to reduce the impact on public health. *Environ. Res.* **2017**, *155*, 219–227. [CrossRef]

52. Cariñanos, P.; Marinangeli, F. An Updated proposal of the Potential Allergenicity of 150 ornamental trees and shrubs in Mediterranean Cities. *Urban For. Urban Green.* **2021**, *63*, 127218. [CrossRef]

53. Nelson, N.L.; Shih, R.P.; Ballick, M.J. *Handbook of Poisonous and Injurious Plants*; Springer: New York, NY, USA; The New York Botanical Garden: New York, NY, USA, 2007.

54. Wagstaff, J. *International Poisonous Plants Checklist. An Evidence-Based Reference*; CRC Press/Taylor and Francis Group: Boca Raton, FL, USA, 2007.

55. Bayón, A.; Vilá, M. Horizon scanning to identify invasion risk of ornamental plants marketed in Spain. *NeoBiota* **2019**, *52*, 47–68. [CrossRef]

56. Sanz Elorza, M.; Dana Sanchez, E.; Sobrino Vesperias, E. *Plantas Alóctonas Invasoras de España*; Ministerio de Medio Ambiente, Dirección General para la Biodiversidad, Gobierno de España: Madrid, Spain, 2004.

57. Hettinger, N. Exotic species, naturalisation, and biological nativism. *Environ. Values* **2011**, *10*, 193–224. [CrossRef]

58. Bodilis, P.; Francour, P.; Langar, H.; El Asmi, S. *Non-Native Species in the Mediterranean: What, When, How and Why?* UNEP-MAP-RAC/SPA; RA/SPA: Tunis, Tunisia, 2011.

59. Singh, R.; Singh, M.P.; Sing, A.P. Ozone forming potential of tropical plant species of the Vidarbha region of Maharashtra state of India. *Urban For. Urban Green.* **2014**, *13*, 814–820. [CrossRef]

60. Karl, M.; Guenther, A.; Köble, R.; Leip, A.; Seufert, G. A new European plant-specific emission inventory of biogenic volatile organic compounds for use in atmospheric transport model. *Biogeosciences* **2009**, *6*, 1059–1087. [CrossRef]

61. Güngor, S.; Demir, M.; Polat, A.T. A Research on Accessibility of urban parks by disabled users. *Int. J. Res. Soc. Sci.* **2016**, *6*, 578–594.

62. Tarim, A. Position of water elements as an urban furniture. *Int. J. Res. Clin. Metall. Civ. Eng.* **2017**, *4*, 43–44.

63. Zuo, K.; Wei, L.; Cong, Y. Exploration of Natural Playgrounds in Urban Parks: Promoting Childrens's Health. *Urban Reg. Plan.* **2020**, *5*, 122–131. [CrossRef]

64. Mahrour, A.M.; Moustafa, Y.M.; El-Ela, M.A.A. Physical characteristics and perceived security in urban parks: Investigation in the Egyptian context. *Ain Shams Eng. J.* **2018**, *9*, 3055–3066. [CrossRef]
65. Chow, H.W. Outdoor fitness equipment in parks: A qualitative study from older adults. *BMC Public Health* **2013**, *13*, 1216. [CrossRef]
66. Voigt, A.; Kabish, N.; Wurster, D.; Haase, D.; Breuste, J. Structural Diversity: A multi-dimensional approach to assess recreational services in urban parks. *AMBIO* **2014**, *43*, 480–491. [CrossRef]
67. Granz, G.; Boland, M. Defining the Sustainable Park: A Fith Model for Urban Parks. *Landsc. J.* **2004**, *23*, 102–120.
68. i-Tree User's Manual. Tools for Assessing and Managing Community Forests. Software Suite v6. 2016. Available online: https://www.itreetools.org/resources/manuals/i-Tree%20Species%20Users%20Manual.pdf (accessed on 15 July 2021).
69. Moreno-García, M.C. The microclimate effect of Green Infrastructure (GI) in a Mediterranean city: The case of the urban park of Ciutadella (Barcelona, Spain). *Agric. Urban For.* **2019**, *45*, 100–108.
70. Zhou, W.; Wang, J.; Cadenasso, M.L. Effects of the Spatial configuration of trees on urban heat mitigation: A comparative study. *Remote. Sens. Environ.* **2017**, *195*, 1–12. [CrossRef]
71. Flinker, P. *The Need to Reduce Impervious Cover to Prevent Flooding and Protect Water Quality*; Department of Environmental Management: Providence, RI, USA, 2010.
72. Wu, D.; Wang, Y.; Fan, C.H.; Xia, B. Thermal environment affects and interactions of reservoir and forests as urban green infrastructures. *Ecol. Indic.* **2018**, *91*, 657–663. [CrossRef]
73. Zuvela-Aloize, M.; Koch, R.; Bucholz, S.; Früh, B. Modelling the potential of green and blue infrastructure to reduce urban heat load in the city of Vienna. *Clim. Chang.* **2016**, *135*, 425–438. [CrossRef]
74. Alonso-Martínez, P.; Castro, M.C.; Pinto-Gomes, C. Flowering meadows, a biodiverse alternative to lawns in Mediterranean urban spaces. In *The Garden as a Lab. Where Cultural and Ecological Systems Meet in the Mediterranean Context*; Conference Book of Abstracts; Universidade de Évora: Évora, Portugal, 2014; pp. 13–147.
75. Forestry Commission. *Putting the Green in the Grey. Creating Sustainable Grey Infrastructure. A Guide for Developers, Planners, and Project Managers*; Forestry Commission: Bristol, UK, 2009.
76. Segnestam, L. *Indicators of Environment and Sustainable Development: Theories and Practical Experience*; Environmental Economics Series 89; The World Bank Environment Department: Washington, DC, USA, 2002.
77. Escobedo, F.J.; Kroeger, T.; Wagner, J.E. Urban forests and pollution mitigation: Analyzing ecosystem services and disservices. *Environ. Pollut.* **2011**, *159*, 2078–2087. [CrossRef]
78. Mediterranean Experts on Climate and Environmental Change (MedECC). Risks Associated to Climate and Environmental Changes in the Mediterranean Region. A Preliminary Assessment by the MedECC Science-Policy Interface. Available online: https://ufmsecretariat.org/wp-content/uploads/2019/10/MedECC-Booklet_EN_WEB.pdf (accessed on 6 August 2021).
79. Maes, J.; Zulian, G.; Günther, S.; Thijssen, M.; Raynal, J. *Enhancing Resilience of Urban Ecosystems through Green Infrastructure*; Final Report, EUR 29630 EN.; Publications Office of the European Union: Luxembourg, 2019. [CrossRef]
80. Borelli, S.; Conigliaro, M.; Pineda, F. Urban Forests in the global Context. *Unasylva* **2018**, *69*, 3–10.
81. Martí, P.; Serrano-Estrada, L.; Nolasco-Cirugeda, A. Using locative social media and urban cartographies to identify and locate succesful urban plazas. *Cities* **2017**, *64*, 66–78. [CrossRef]
82. Oktay, D. Smart and Sustainable Urbanism with Identity in Mediterranean Cities. In *Green Buildings and Renewable Energy. Innovative Renewable Energy*; Sayigh, A., Ed.; Springer: Cham, Switzerland, 2020.
83. Cariñanos, P.; Salbitano, F.; Conigliaro, M. Foretst and Cities: Forests-Based Solutions in Urban Areas. *Forèt Méditerranéennee* **2019**, *3*, 261–268.
84. Cariñanos, P.; Casares-Porcel, M. Urban Green Zones and Related Pollen Allergy: A review. Some guidelines for designing spaces with low allergy impact. *Landsc. Urban Plan.* **2011**, *101*, 205–214. [CrossRef]
85. Ren, Y.; Qu, Z.; Du, Y.; Xu, R.; Ma, D.; Yang, G.; Shi, Y.; Fan, X.; Tani, A.; Guo, P.; et al. Air quality and health effects of biogenic volatile organic compounds emissions from urban green spaces and the mitigation strategies. *Environ. Pollut.* **2017**, *230*, 849–861. [CrossRef]
86. Grote, R.; Samson, R.; Alonso, R.; Amorim, J.H.; Cariñanos, P.; Churkina, G.; Fares, S.; Thiec, D.L.; Niinemets, Ü.; Mikkelsen, T.N.; et al. Functional traits of Urban trees: Air pollution mitigation potential. *Front. Ecol. Environ.* **2016**, *14*, 543–550. [CrossRef]
87. Manescu, C.R.; Dpbrescu, E.; Nujnoi, S.; Toma, F.; Petra, S. Toxic plants species in parks located in city centre of Bucharest. Scientific Papers. Series B. *Horticulture* **2019**, *63*, 529–534.
88. Constán-Navas, S.; Bonet, A.; Pastor, E.; Lledó, M.J. Long-term control of the invasive tree *Ailanthus altissima*: Insights from Mediterranean protected forests. *For. Ecol. Manag.* **2020**, *260*, 1058–1064. [CrossRef]
89. Abdelhamid, M.M.; Elfakharany, M.M. Improving urban park usability in developing countries: Case-study of Al-Shalalat Park in Alexandria. *Eng. J.* **2020**, *59*, 311–321. [CrossRef]
90. Jiménez-Delgado, A.; Lloret, J. *Health, Wellebing and Sustainability in the Mediterranean City. Interdisciplinary Perspectives. Routledge Studies in Urbanism and the City*; Taylor and Francis Group: Boca Raton, FL, USA, 2019.
91. Transnational Institute (Amsterdam) and Latin American Council of Social Sciences. *Public Water and COVID-Dark Clouds and Silver Linnings*; Municipal Service Project; McDonald, D.A., Sprouk, S.J., Chaves, D., Eds.; Transnational Institute (Amsterdam) and Latin American Council of Social Sciences: Amsterdam, The Netherlands, 2021.

92. Ayala-Azcárraga, C.; Díaz, D.; Zambrano, L. Characteristics of urban parks and their relation to well-being. *Landsc. Urban Plan.* **2019**, *189*, 27–35. [CrossRef]
93. McComack, G.R.; Rpck, M.; Toohey, A.M.; Hignell, D. Characteristics of urban parks associated with park use and physical activity: A Review of qualitative research. *Health Place* **2010**, *16*, 712–726. [CrossRef] [PubMed]
94. Dempsey, N.; Burton, M. Defining place-keeping: The long-term management of public spaces. *Urban For. Urban Green.* **2012**, *11*, 11–20. [CrossRef]
95. Duinker, P.N.; Lehvävirta, S.; Busse Nielsen, A.; Toni, S.A. Urban woodlands, and their management. In *Routledge Handbook of Urban Forestry*; Ferrini, F., Konijnendijk, C.C., Fini, A., Eds.; Routledge, Taylor and Francis Group: Boca Raton, FL, USA, 2019.
96. Maruthaveeran, S.; van den Bosch, C.C.K. A socio-ecological exploration of fear of crime in urban green spaces. A systematic review. *Urban For. Urban Green.* **2014**, *13*, 1–18.
97. Tandogan, O.; ILhan, B.S. Fear of crime in public spaces: From the view of women living in cities. *Procedia Eng.* **2016**, *161*, 2011–2018. [CrossRef]
98. Toribio Marín, C. The fourfold Water Garden, a Renaissance Invention. *Gard. Landsc.* **2016**, *4*, 1–16. [CrossRef]
99. Pérez-Igualada, J. Water-sensitive strategies in the new urban parks in Valencia: The Agricultural Mediterranean paradigm as a pattern for landscape management and design. *Int. J. Sustain. Dev. Plan.* **2016**, *9*, 97–106. [CrossRef]
100. Hua, L.; Zhang, X.; Nie, Q.; Sun, F.; Tang, L. The impacts of the Expansion of Urban Impervious Surfaces on Urban Heat Islands in Coastal City in China. *Sustainability* **2020**, *12*, 475. [CrossRef]
101. AEMET-OECC. *Cambio Climático: Calentamiento Global de 1.5 °C*; Agencia Estatal de Meteorología y Oficina Española de Cambio Climático, Ministerio para la Transición Ecológica: Madrid, Spain, 2018.
102. Ferguson, B.K. Permeable pavements in Liveable, Sustainable Cities. *CityGreen* **2018**, *5*, 34–39. [CrossRef]
103. Good Gardening Practices in Barcelona. Conserving and Improving Biodiversity. *Ayuntamiento de Barcelona*. 2016. Available online: https://ajuntament.barcelona.cat/ecologiaurbana/sites/default/files/Bones-practiques-jardineria-2016-ENG.pdf (accessed on 10 September 2021).
104. Jiménez-Alfaro, B.; Frischi, S.; Stolz, J.; Gálvez, C. Native plants for greening Mediterranean agroecosystems. *Nat. Plants* **2020**, *6*, 209–214. [CrossRef]
105. Nowak, D.J.; Heisler, G.M. *Air Quality Effects of Urban Trees and Parks*; Research Series; National Recreation and Park Association: Ashburn, VA, USA, 2010.
106. Li, G.; Wan, L.; Cui, M.; Wu, B.; Zhou, J. Influence of Canopy Interception and Rainfall Kinetic Energy on Soil under Forests. *Forests* **2019**, *10*, 509. [CrossRef]
107. Peel, M.C.; Finlayson, B.L.; McMahon, T.A. Updated world map of the Köppen-Geiger climate classification. *Hydrol. Earth Syst. Sci.* **2007**, *11*, 1633–1644. [CrossRef]
108. Pantovou, K.G.; Jacovides, C.P.; Nikolopoulo, G.K. Data on solar sunburning ultraviolet (UVB) radiation at an urban Mediterranean climate. *Data Brief* **2017**, *11*, 597–600. [CrossRef] [PubMed]
109. Li, X.; Ratti, C.; Seiferling, I. Quantifying the shade provision of street trees in urban landscape: A case study in Boston, USA, using Google Street view. *Landsc. Urban Plan.* **2018**, *169*, 81–91. [CrossRef]
110. Oliveira, S.; Vaz, T.; Andrade, H. Perception of thermal comfort by users of urban green areas in Lisbon. *Finisterra* **2014**, *49*, 113–131. [CrossRef]

Article

Response of Spatio-Temporal Differentiation Characteristics of Habitat Quality to Land Surface Temperature in a Fast Urbanized City

Yongge Hu [1,†], Enkai Xu [1,†], Gunwoo Kim [2,*], Chang Liu [1] and Guohang Tian [1,*]

[1] Department of Landscape Architecture, College of Landscape Architecture and Art, Henan Agricultural University, Zhengzhou 450002, China; huyonggec@gmail.com (Y.H.); xek1206@henau.edu.cn (E.X.); cliu229@uic.edu (C.L.)

[2] Graduate School of Urban Studies, Hanyang University, 222 Wangsimni-ro, Seongdong-gu, Seoul 04763, Korea

[*] Correspondence: gwkim1@hanyang.ac.kr (G.K.); tgh@henau.edu.cn (G.T.); Tel.: +82-2-2220-0274 (G.K.); +86-0371-6355-8098 (G.T.)

[†] These authors contributed to this work equally and should be regarded as co-first authors.

Abstract: The degradation and loss of global urban habitat and biodiversity have been extensively studied as a global issue. Urban heat islands caused by abnormal land surface temperature (LST) have been shown to be the main reason for this problem. With the accelerated urbanization process and the increasing possibility of abnormal temperatures in Zhengzhou, China, more and more creatures cannot adapt and survive in urban habitats, including humans; therefore, Zhengzhou was selected as the study area. The purpose of this study is to explore the response of urban habitat quality to LST, which provides a basis for the scientific protection of urban habitat and biodiversity in Zhengzhou from the perspective of alleviating heat island effect. We used the InVEST-Habitat Quality model to calculate the urban habitat quality, combined with GIS spatial statistics and bivariate spatial autocorrelation analysis, to explore the response of habitat quality to LST. The results show the following: (1) From 2000 to 2015, the mean value of urban habitat quality gradually decreased from 0.361 to 0.304, showing a downward trend as a whole. (2) There was an obvious gradient effect between habitat quality and LST. Habitat quality's high values were distributed in the central and northern built-up area and low values were distributed in the high-altitude western forest habitat and northern water habitat. However, the distribution of LST gradient values were opposite to the habitat quality to a great extent. (3) There were four agglomeration types between LST and habitat quality at specific spatial locations: the high-high type was scattered mainly in the western part of the study area and in the northern region; the high-low type was mainly distributed in the densely populated and actively constructed central areas; the low-low type was mainly distributed in the urban-rural intersections and small and medium-sized rural settlements; and the low-high type was mainly distributed in the western mountainous hills and the northern waters.

Keywords: urban biodiversity; urban habitat quality; InVEST model; land surface temperature; Moran's *I*; Zhengzhou

Citation: Hu, Y.; Xu, E.; Kim, G.; Liu, C.; Tian, G. Response of Spatio-Temporal Differentiation Characteristics of Habitat Quality to Land Surface Temperature in a Fast Urbanized City. *Forests* **2021**, *12*, 1668. https://doi.org/10.3390/f12121668

Academic Editor: Elisabetta Salvatori

Received: 1 November 2021
Accepted: 26 November 2021
Published: 30 November 2021

Publisher's Note: MDPI stays neutral with regard to jurisdictional claims in published maps and institutional affiliations.

1. Introduction

Urban biodiversity is the driving force of sustainable urban development, and it is of great significance for maintaining a balanced, sustainable human settlement environment [1]. Biodiversity loss, habitat degradation, and the urban heat island effect are generally considered to be intertwined urban problems [2]. Urban heat islands have led to continuous changes in urban species abundance, distribution patterns, and energy flow and material cycling; further, some organisms have been forced to migrate or have even become extinct [3–5].

Land surface temperature (LST) is an important indicator that reflects the intensity of urban heat islands, and characterizes the energy balance and climate change on the earth's surface [6,7]. It has a significant impact on the migration of surface matter and energy conversion. The use of thermal infrared remote sensing can obtain a wide range of LST information compared with traditional measurement methods. It is fast and convenient and has a large measurement range, continuous information, and so on, which can be used to measure the characteristics of regional climate change [8,9]. Although the land area of physically urbanized cities only accounts for 3% of the total land surface area of the earth, the ecological footprint of cities is usually hundreds of times their area [10]. The rise in LST caused by urban activities directly impacts the time and energy that a species must invest in maintaining an optimal body temperature [11]. World Cities Report 2020 from UN-Habitat's pointed out that the world will be further urbanized in the next ten years. The proportion of the global urban population will increase from the current 56.2% to 60.4% in 2030, which means urban construction activities will be further expanded and the urban LST differentiation may be more prominent.

Habitat quality refers to the ability of a certain space to provide resources and ensure survival and reproduction for species, which can be regarded as an important characterization and reflection of regional biodiversity [12]. The "Global Biodiversity Outlook 5" report (https://www.cbd.int/gbo/, accessed on 11 August 2021) reveals that the vertebrate population has fallen by more than two-thirds on average since 1970, especially in built-up areas outside nature reserves, where the trend of biodiversity loss has not been effectively curbed.

Improving the habitat quality on which organisms depend for survival is a basic way to protect biodiversity [13–15]. Generally, habitat quality decreases with the intensity of human land use [16,17]. The higher the habitat quality, the higher the level of biodiversity and the higher the service capacity of the ecosystem. In the 1980s, based on the experimental data of field surveys focusing on the geographical differentiation characteristics of the habitat conditions of species in a specific area, the evaluation results were generally accurate [18,19]. However, the operation is complex, the cycle is long, and it is not easy to promote applications and carry out comparative studies.

After the 21st century, as the integration of RS (remote sensing), GIS (geography information systems) and GPS (global positioning systems) becomes stronger, it is possible to quantitatively, visually, and finely analyze and assess the habitat changes in the spatio-temporal scale of large- and medium-scale. At this stage, the research and application of ecological models such as SolVES (Social Values for Ecosystem Services) [20], ARIES (artificial intelligence for ecosystem services) [21], and InVEST (Integrated Valuation of Ecosystem Services and Tradeoffs) have become the most important in the field of urban ecology. Among many research hotspots of ecosystem service assessment, the InVEST model is the most widely used [22,23]. It was jointly developed by Stanford University, The Nature Conservancy (TNC), and the World Wide Fund for Nature (WWF) to evaluate and weigh ecosystem services. Habitat Quality module is based on the principle of habitat stress to quickly assess the quality of regional habitats. The required data can replace complex methods, such as species surveys, because they are easy to obtain and have a wide range of applications.

At present, research on the characteristics of the spatio-temporal differentiation of habitat quality mainly focuses on grassland [24], wetland [25], watershed [26,27], and forest [28] habitats, among others; research on the driving factors of habitat quality mostly focuses on land use change [29,30], landscape pattern [31,32], topography [33], urban expansion [34], road network [35,36], and species invasion [37]. However, there is a lack of research on the response mechanism of urban habitat quality to LST, especially in the open and complex mega-system metropolis. The purpose of this study is to explore the spatio-temporal mechanism of urban habitat for LST, and adjust the LST to improve the quality of urban habitat. Zhengzhou is a typical agglomeration city under the background of urbanization and an important comprehensive hub in central China. The city is currently facing stricter resource and environmental restrictions, which have a negative impact on the urban habitat and biodiversity. In view of this, Zhengzhou was selected as the research object. Based on remote sensing and land use data from 2000 to 2015, we assessed the spatio-temporal differentiation characteristics of Zhengzhou's habitat quality using the InVEST-Habitat Quality model. GIS spatial statistics and GeoDa bivariate spatial correlation were used to analyze the response of habitat quality to LST on the urban area. The results were expected to provide data support for biodiversity conservation and land use planning in megacities.

2. Study Area

Zhengzhou (34°16′–34°58′ N, 112°42′–114°14′ E), the capital of Henan Province, is located in the central hinterland of China in the eastern section of the Qinling Mountains, the transitional zone between the second-level and the third-level landform steps in China, with a total area of 7576 km^2 and a population of more than 12 million (Figure 1). The terrain is high in the southwest and low in the northeast, descending in a stepped manner. From west to east, it gradually transits from middle and low mountains to tectonic denudation hills, loess hills, slope plains, and alluvial plains, forming a relatively complete geomorphic sequence.

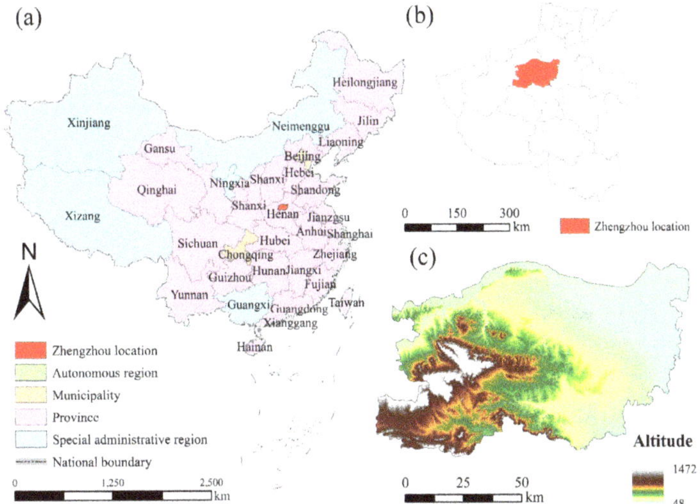

Figure 1. Geographic location (**a**), administrative district (**b**), and topography (**c**) of Zhengzhou region.

The city's main habitats include cultivated land, woodland, grassland, water, construction land, and unused land. Woodland is mainly distributed in the western and southwestern mountains, including Mang Mountain, Song Mountain, Fu Xi Mountain, and Big Bear Mountain, which constitute an ecological barrier in the west of Zhengzhou; the northern Yellow River is the main water area that flows through Zhengzhou, and its banks are located in the middle of the three major migration channels of migratory birds in China. It is an important migratory habitat and wintering place for migratory birds. It is one of the key areas of biodiversity distribution in China and has important ecological value. Cultivated land and grassland are mainly distributed in the southern and eastern plains, and construction land is mainly distributed in the central urban area of Zhengzhou and the built-up areas of cities under its jurisdiction.

3. Data Sources and Methods

3.1. Data Used

The workflow steps in our analysis are shown in Figure 2. The principle of the InVEST-Habitat Quality model is to combine land-use and land-cover change (LUCC) and threat factors to generate a habitat quality map to reflect the status of regional biodiversity. The interpretation data of land use and cover in 2000, 2005, 2010, and 2015 required to calculate habitat quality are all sourced from the Data Center for Resources and Environmental Sciences, Chinese Academy of Sciences (RESDC) [38,39]. The grid resolution is 30 m × 30 m. The Landsat series of remote sensing images required to retrieve LST are provided by the United States Geological Survey (https://earthexplorer.usgs.gov/, accessed on 6 January 2021). To maintain consistency of the data source, images of the same month are selected to minimize the effects of seasonal variation in solar radiation. The dates of the selected images are 8 June 2000 (Landsat 7), 22 June 2005 (Landsat 7), 20 June 2010 (Landsat 7), and 10 June 2015 (Landsat 8). To reduce the difference in topography, illumination, and atmosphere of different time phase images, radiation correction and atmospheric correction were carried out on the four phase images, and the data processing was carried out in ArcGIS 10.2 and ENVI 5.3.

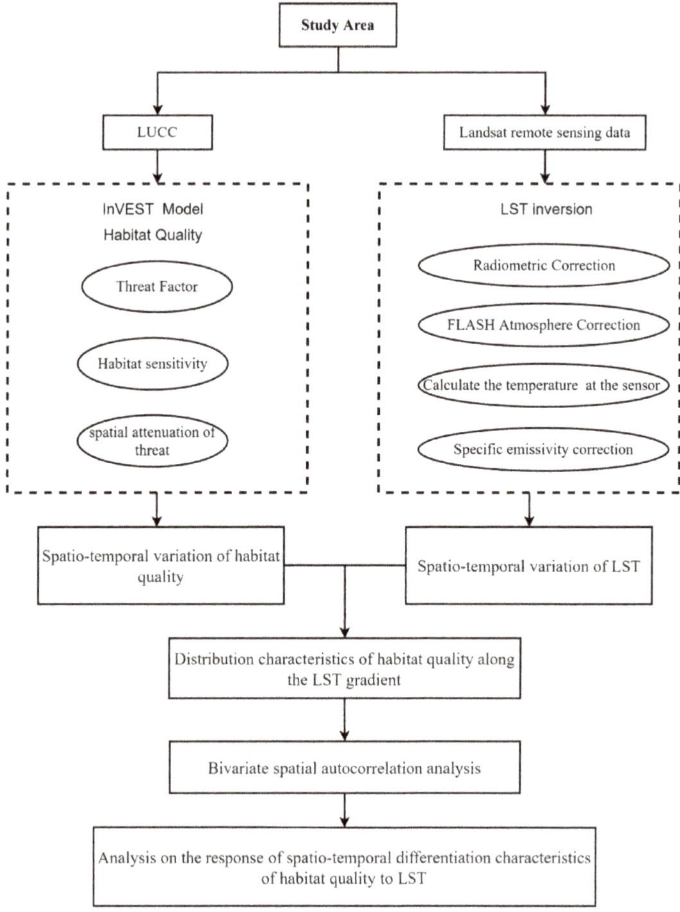

Figure 2. Workflow steps in our analysis.

3.2. Habitat Quality Calculation

The InVEST-Habitat Quality runs using raster data, where each cell in the raster is assigned an LUCC class. The user defines which LUCC types can provide habitat for the conservation objective. Generally, a relative habitat suitability value can be assigned to an LULC type ranging from 0 to 1. The higher the value, the higher the suitability for the survival and reproduction of individual organisms or populations.

In addition to a map of LUCC and data that relates LUCC to habitat suitability, the model also requires data on habitat threat density and its effects on habitat quality. Each threat source needs to be mapped on a raster grid and all mapped threats should be measured in the same scale and metric. It is worth emphasizing that the impact of threats on habitat in a grid cell is dominated by two main factors. One is the threat sources selected in specific research and their relative impact, the other is the distance between habitat and the threat source and the impact of the threat across space.

Based on the basic situation of the study area, the InVEST User's Guide and previous research results [32,40–43], we defined paddy field, dry land, built-up land, rural residential areas, and other construction land (quarries, mines, traffic land, etc.) as threat sources, and assigned values to the sensitivity of each habitat to threat factors (Tables 1 and 2). The calculation is as follows:

$$D_{xj} = \sum_{r=1}^{R}\sum_{y=1}^{Y_r}\left(\frac{\omega_r}{\sum_{r=1}^{R}\omega_r}\right)r_y i_{rxy}\beta_x S_{jr} \tag{1}$$

Table 1. Habitat types and sensitivity of habitat types to each threat.

Habitat Type		Habitat Suitability	Threat Factor				
Main Category	Subcategory		Paddy Field	Dry Land	Built-Up Land	Rural Residential Land	Other Construction Land
Cultivated land	Paddy field	0.4	0	1	0.5	0.3	0.4
	Dry land	0.3	1	0	0.5	0.3	0.4
Forest	Woodland	1	0.8	0.9	0.9	0.8	0.8
	Shrubwood	0.7	0.4	0.5	0.6	0.4	0.5
	Sparse woodland	0.6	0.8	0.9	0.5	0.7	0.8
	Other woodland	0.5	0.9	1	0.5	0.7	0.8
Grassland	High coverage grassland	0.7	0.4	0.5	0.8	0.45	0.7
	Medium coverage grassland	0.5	0.5	0.6	0.7	0.5	0.6
	Low coverage grassland	0.3	0.7	0.9	0.6	0.55	0.5
Water	River and canal	1	0.7	0.5	0.7	0.6	0.6
	Lake	0.9	0.7	0.5	0.7	0.6	0.6
	Reservoir and pond	0.8	0.8	0.4	0.8	0.7	0.7
	Bottomland	0.6	0	0	0.9	0.8	0.7
Construction land	Built-up land	0	0	0	0	0	0
	Rural residential land	0	0	0	0	0	0
	Other construction land	0	0	0	0	0	0
Unused land	Sandy land	0.1	0.1	0.1	0.2	0.1	0.1
	Saline-alkali land	0.1	0.1	0.1	0.2	0.1	0.1
	Marshland	0.6	0.5	0.6	0.8	0.6	0.5
	Bare soil land	0	0	0	0	0	0
	Bare rock land	0	0	0	0	0	0

Table 2. Threats' weight and the maximum distance of influence.

Threat Source	Relative Weight	Maximum Influence Distance (km)	Spatial Attenuation Function
Paddy field	0.4	4	Exponential
Dry land	0.3	4	Exponential
Built-up land	1	8	Exponential
Rural residential land	0.6	6	Exponential
Other construction land	0.8	5	Exponential

In Equation (1), D_{xj} represents the degree of habitat degradation; R represents the total threat sources and r represents one of the threat sources; y represents the grid unit of the r threat layer; Y_r refers to the sum of the grids on the r threat layer; ω_r represents the weight of threat source r; r_y is used to judge whether grid y is the source of threat source r; β_x represents the protection of society, law, etc., which is not taken into consideration in this research. The accessibility level from threat source r to grid x, S_{jr} represents the sensitivity of habitat type j to threat source r. i_{rxy} represents the distance impact function of threat source r in the habitat of grid x on grid y. Assuming that the impact of the threat source on the habitat decays exponentially, the calculation is as follows:

$$i_{rxy} = exp\left[1 - \left(\frac{2.99}{d_{rmax}}\right)d_{xy}\right] \tag{2}$$

In Equation (2), d_{xy} represents the linear distance between grid x and y, and d_{rmax} represents the maximum impact distance of threat source r.

Therefore, according to the investment model's calculation principle of habitat quality, the calculation is as follows:

$$Q_{xj} = H_j\left[1 - \left(\frac{D_{xj}^Z}{D_{xj}^Z + k^Z}\right)\right] \tag{3}$$

In Equation (3), Q_{xj} represents habitat quality; H_j represents the habitat suitability of land use type j; k is the half-saturation constant, which is usually set to half of the highest degradation value; Z is the default parameter of the model, $Z = 2.5$.

3.3. LST Inversion Calculation

As a key parameter in many basic disciplines and applications, LST can provide information about spatio-temporal variation of the surface energy balance. The atmospheric correction method adopted in this study is mainly based on the composition of the thermal radiation intensity observed by the remote sensor on the satellite to calculate LST [44]. The related calculation is as follows:

$$L = gain \times DN + bias \tag{4}$$

In Equation (4), L is the radiation value of the pixel in the ETM+ thermal infrared band at the sensor, DN is the gray value of the pixel, and $gain$ and $bias$ are the gain and bias values of the thermal infrared band, respectively, which can be obtained from the image header file.

$$T = K_2 \; / \; \ln(K_1/L + 1) \tag{5}$$

In Equation (5), T is the temperature value at the sensor; K_1 and K_2 are calibration parameters, Landsat 7: $K_1 = 666.093$ W/(m²·sr·µm), $K_2 = 1282.710$ K; Landsat 8: $K_1 = 774.885$ W/(m²·sr·µm), $K_2 = 1321.079$ K. The sensor temperature (T) must be corrected for specific emissivity to estimate the final LST.

$$LST \; = \; T \; / \; [1 + (\lambda T/\rho)\ln \varepsilon]. \tag{6}$$

In Equation (6), λ is the center wavelength of the ETM+ 6 band ($\lambda = 11.335$ µm); $\rho = 1.438 \times 10^{-2}$ m·K; and ε is the surface emissivity, which is estimated using the *NDVI* threshold method [45].

3.4. Bivariate Moran's I Calculation

After preliminary analysis, we found that the spatial numerical distribution of habitat quality and LST were very similar, and both had a certain degree of spatial autocorrelation. Bivariate spatial autocorrelation has high applicability and effectiveness in describing the spatial correlation and dependence characteristics of two geographic elements. At present, this method has not been found to be used to explore the spatial relationship between habitat quality and LST. Therefore, we have innovatively attempt to adopt this method.

Bivariate Moran's *I* is an extension and expansion based on Moran's *I* index, which measures the correlation between the attribute values of spatial units and other attribute values in adjacent spaces [46,47]. It can be used as an effective method to analyze the correlation characteristics between urban habitat quality distribution and LST. Bivariate Moran's *I* is divided into two levels: global Moran's *I* and local Moran's *I*. The calculation formula is as follows:

$$I_{ab} = \left(\frac{X_{ma} - \overline{X}_a}{\delta_a} \right) \left(\frac{X_{ob} - \overline{X}_b}{\delta_b} \right) \Sigma_{j=1}^{n} W_{mo} \tag{7}$$

In Equation (7), X_{ma} is the value of the variable *a* of the spatial unit *m*, X_{ob} is the value of the variable *b* of the spatial unit *o*, $\overline{X}_a$ and $\overline{X}_b$ are the mean values of *a* and *b*, respectively, and δ_a and δ_b are the variances of *a* and *b*; W_{mo} is the spatial weight matrix between unit *m* and *o*. I_{ab} is the Moran's *I* statistic and its value is between (−1, 1): less than 0 means negative correlation, equal to 0 means no correlation, and greater than 0 means positive correlation. Data processing is done in GeoDa 1.6.7.

4. Results

4.1. Spatio-Temporal Variation of Habitat Quality

Using the isometric classification method of ArcGIS, the calculation results of habitat quality index were divided into five intervals: 0.00–0.20, 0.20–0.40, 0.40–0.60, 0.60–0.80, and 0.80–1, which represent I, II, III, IV, and V (from low to high), respectively. The differences in habitat quality and the evolution law of spatio-temporal pattern in Zhengzhou were further analyzed (Figures 3 and 4).

From the perspective of the time scale, the mean habitat quality of Zhengzhou gradually decreased from 0.361 to 0.304 from 2000 to 2015, with II accounting for the highest proportion of the total area, with an annual mean of 62.94%. Especially, the continuum of habitats located at the intersection of the urban built-up area was gradually fragmented to form scattered and isolated island-like habitats or habitat fragments.

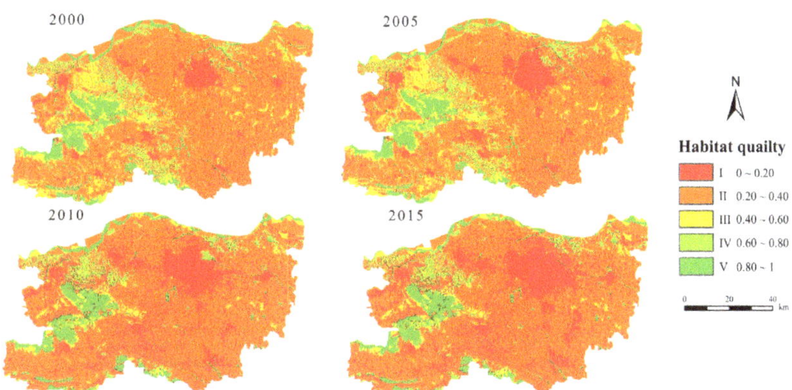

Figure 3. Spatial distribution of habitat quality during the different periods from 2000 to 2015.

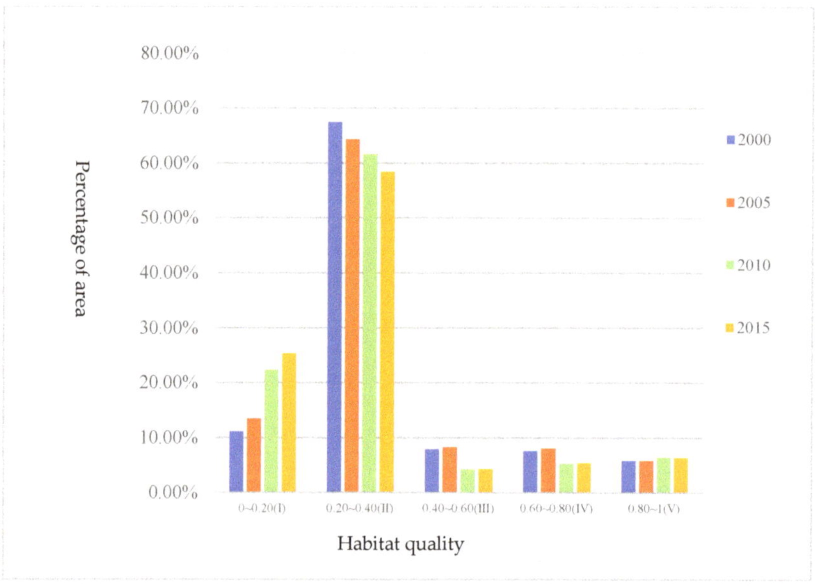

Figure 4. Percentage of the five levels of habitat quality area from 2000 to 2015.

From the perspective of the spatial pattern, the spatial distribution of habitat quality in Zhengzhou is significantly different. II is the most widely distributed, mainly in the low-elevation plains in the east, where human farming is frequent, the ecosystem is single, the vegetation coverage rate is low, and the ecological environment is relatively fragile. I is concentrated in urban built-up areas; the land use structure in these places is dominated by construction land, human activities are disturbed, and biodiversity is low. III, IV, and V are mainly distributed in the high-altitude western mountain and hilly areas, including Fuxi Mountain, Changshou Mountain, Qinglong Mountain, and Song Mountain. These regions have a sparse population with very high green space coverage fraction ratio. Moreover, several ecological protection policies have been implemented, making the habitats there more suitable for many kinds of animals and plants. Further, there are also some IV or higher value habitats in the northern part of the city, thanks to the abundant waters and wetlands—resources suitable for the survival of aquatic animals, plants, and birds.

4.2. Spatio-Temporal Variation of LST

To eliminate the influence of dimension and magnitude on the final result, this study uses the range standardization method to process the LST results of Zhengzhou; the value range was (0, 1). Using ArcGIS's equal interval classification method, LST was divided into 10 gradients from low to high (Figure 5). In terms of time scale, LST dropped slightly from 2000 to 2005, rose markedly in 2010, and dropped slightly in 2015. The results show that in different stages of Zhengzhou's transformation from traditional industry and agriculture to modern service industry, its overall LST is in a dynamic change process.

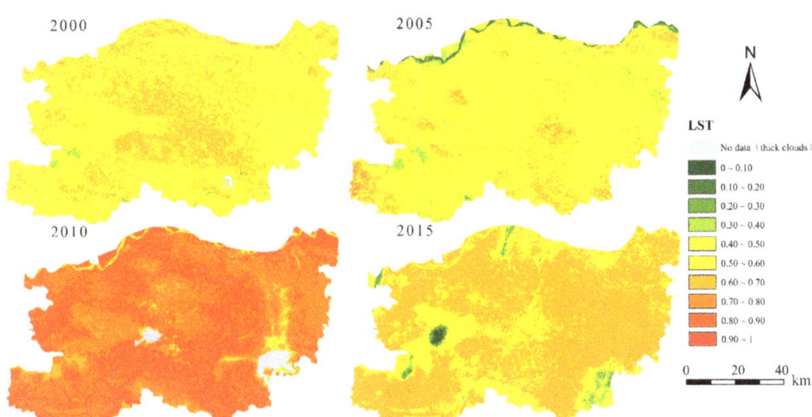

Figure 5. Spatial distribution of LST during the different periods from 2000 to 2015.

From the spatial distribution perspective, high gradients (8, 9, 10) are distributed in the central city and southern airport area which indicates that the high island effect is mainly affected by human activities; low gradients (1, 2, 3, 4) showed an obvious spatial agglomeration effect in the western mountainous areas and northern waters which indicates that natural habitats have a significant effect on alleviating the heat island effect and are easy to become the center of the cold island effect. The medium gradient (5, 6, 7, 8) is widely distributed in Zhengzhou, indicating that the traditional urban heat island had exceed the boundary of the built-up areas and become a regional heat island.

4.3. The Distribution of Habitat Quality along LST Gradients

As seen in Figure 6, there is an apparent numerical gradient effect between habitat quality and LST. From the time scale, the response of habitat quality to LST in the four periods corresponds to four trend lines, which are obviously different. The fluctuation range in 2000 and 2005 is larger than that in 2010 and 2015.

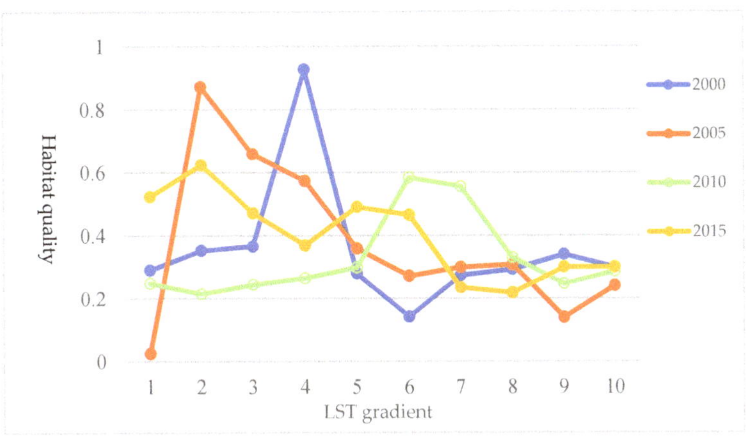

Figure 6. Distribution of habitat quality along LST gradients during the different periods from 2000 to 2015.

From the perspective of spatial scale, high values of habitat quality are concentrated in low gradients (1, 2, 3), and low values are concentrated in high gradients (8, 9, 10), indicating that suitable LST can ensure the stability of habitats. High temperatures can lead to restrictive natural factors such as reduced surface runoff, water shortages, and plants' evapotranspiration, which cause a decline in habitat quality and the ability to support biological communities.

4.4. Bivariate Spatial Autocorrelation Analysis

Using GeoDa software, the bivariate spatial autocorrelation analysis was carried out with LST as the first variable (X) and habitat quality (HQ) as the second variable (Y). Corresponding to different periods in 2000, 2005, 2010 and 2015, the global Moran's *I* were −0.430, −0.313, −0.323, and −0.311, respectively (Figure 7). Randomization 999 was selected in GeoDa for the significance test. The results showed that the *p* values were all 0.001, indicating a significant spatial negative correlation between LST and habitat quality under the confidence of 99.9%; that is, with the increase in LST, habitat quality in Zhengzhou showed a downward trend.

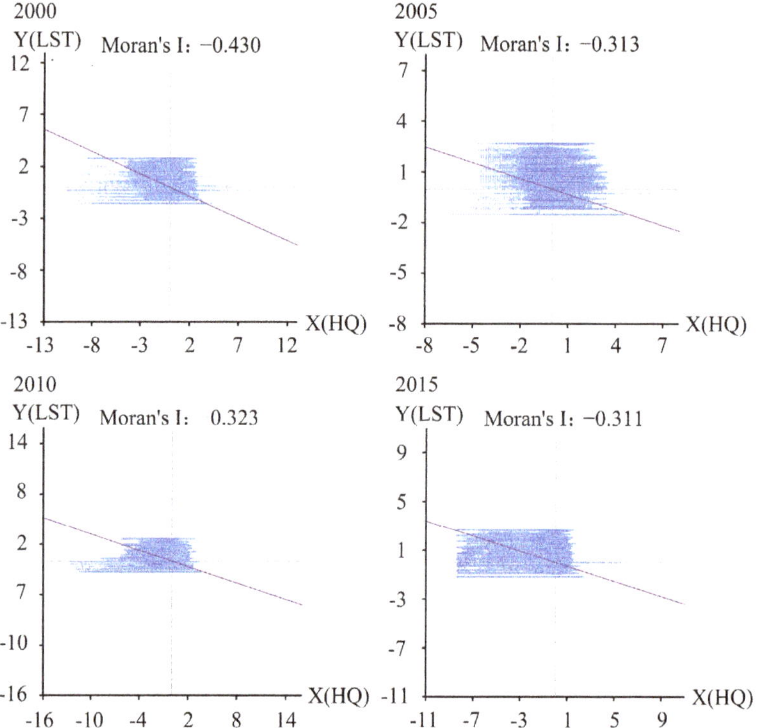

Figure 7. The Moran's *I* scatter diagram of LST and habitat quality during the different periods from 2000 to 2015.

The global Moran's *I* only represents the overall correlation trend of the two variables and cannot reflect the agglomeration differences in specific spatial locations. Therefore, the local Moran's *I* was further calculated, and the results (LST-habitat quality) were divided into four aggregation types: high-high (H-H), low-low (L-L), high-low (H-L), and low-high (L-H). The LISA (Local Indicators of Spatial Association) cluster map of LST and habitat quality clearly show the spatial agglomeration characteristics of regions that have passed the significance test (Figure 8).

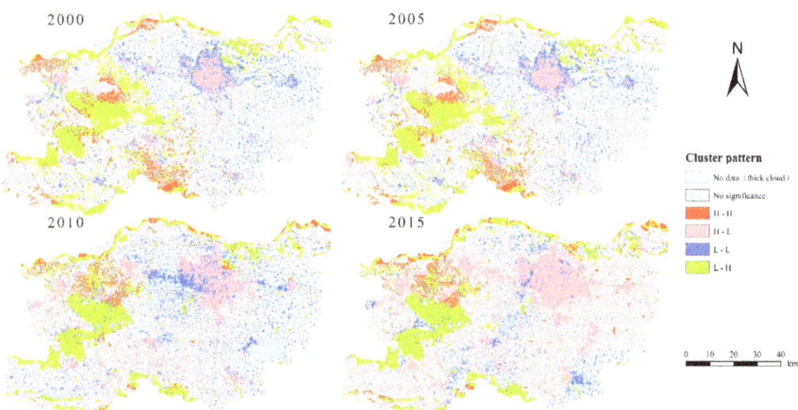

Figure 8. LISA cluster map of LST and habitat quality during the different periods from 2000–2015.

The H-H type was scattered mainly in the western and northern regions of the study area, showing a significant reduction trend from 2000 to 2015, indicating that the rise of a certain LST in some areas in the initial stage can promote the flow of energy and materials in the habitat, but the continuous high LST straightforwardly led to different degrees of structural confusion, dysfunction, and other degradation trends in different habitats. The H-L type was mainly distributed in densely populated and actively built central urban areas, showing a continuous expansion trend, indicating that suburbanization of residential areas is bound to be accompanied by suburban urbanization. The gradual diffusion of manufacturing and industrial agglomeration centers from central urban areas to suburbs is the main reason for the rise of LST. The excessive LST destroys the structure and function of the original habitat. This has resulted in the decline of habitat quality.

The L-L type was mainly distributed in the urban-rural transition zone and small and medium-sized rural residential areas, showing a decreasing trend. The rural areas were dominated by cultivated land habitats in the initial stage of farming. Therefore, the traditional agricultural operation had little impact on the LST, the LST value remained at a low level, and the quality of cultivated land habitat itself was at a low level; thus, it presented a low-low agglomeration trend. However, with the development of mechanized farming technology, artificial heat source increases the LST, and the accumulation type has a great tendency to change to the H-L type.

The L-H type was mainly distributed in the western mountains and hills and the riparian areas of northern waters. It has remained stable in quantity and space for a long time, indicating that the LST in the natural environment was generally lower than that in the human activity area. It was suitable for the growth and reproduction of species.

The results show obvious differences in the spatial correlation of habitat quality and LST in Zhengzhou, and the differentiation results were also evolving with the development of Zhengzhou.

5. Discussion

From the perspective of LST, we analyzed the temporal and spatial differentiation of habitat quality characteristics. It reflects the gradient effect between habitat quality and LST, and reveals the agglomeration characteristics of the two in a specific spatial location.

According to the results of Section 4.2, we can speculate that the intensity and distribution of urban thermal environment is affected by changes in population aggregation, resource consumption, and industrial development, etc. In the initial stage (2000–2005), the rural population around Zhengzhou gathered in built-up areas, and the high temperature agglomerate in small areas, so the overall LST decreased. In the stage of rapid development (2005–2010), with more foreign population and their material demand, a large number of industrial zones had to be built around the built-up area, which were bound to be accompanied by the release of heat energy, so the overall LST rose. In the stage of stable construction (2010–2015), with the strengthening of energy conservation and emission reduction policy, the government was strictly guiding industry to reduce the use of fossil energy, restricting the use of coal and oil, and encouraging subsidies for the development of wind energy, so the overall LST dropped.

According to the results of Section 4.3, we can speculate that in 2000, with increasing construction and production activities, the LST changed suddenly, resulting in a tremendous impact on the quality of urban habitat. After 2010, with the city entering a period of stable development, the government increased the supervision of nature reserves and strictly controlled intensity of urban development. Moreover, the public had also been consciously advocating ecological environment protection and actively responding to global warming. Therefore, The LST gradually returned to the controllable range, which had a weaker impact on the quality of most habitats compared to 2000 and 2005. The prediction was consistent with the variation characteristics of habitat quality in response to LST in Figure 6.

It should be noted that there are still many other possible factors affecting the urban habitat quality, including LUCC, topographic features, meteorology, socio-economic condition, vegetation, etc. The LST is the one factor affecting habitat quality in the region that may interact with other ecological factors. In follow-up studies, the following should be considered: to take grid units of different scales as habitats, to introduce more potential factors in combination with observation stations, and to deeply study the spatio-temporal dynamic evolution law of urban habitat quality to provide a reference for the government to take more accurate ecological management and construction measures to protect urban habitats. In addition, although the LST is easy to obtain through remote sensing inversion, there are random errors in the instantaneous data, which may be affected by the weather and time of the day. As the LST inversion needs the thermal infrared information of ground objects, the shielding of cloudy weather will affect the results. There are also certain differences in the surface temperature during the day, such as evening and noon, and the acquisition time of the existing remote sensing images is not fixed. Further, we will try to set up some field meteorological stations to acquire LST measured data to correct the possible errors caused by both weather and time. In the future, in the construction of urban human settlements and biodiversity protection, we can start from the following aspects:

(1) Carrying out scientific urban planning and building design to reduce the area of the impervious layer. Building green low-carbon buildings to reduce carbon emission reduction in the whole life cycle of buildings is the most ecological and effective means to reduce the "urban heat island" effect and increase urban biodiversity. For example, roof greening can provide a good habitat for some animals, such as butterflies and bees, and provide a certain food source for some birds. It can also use its special habitat ex situ protection box to breed some endangered plant species and increase urban biodiversity. In addition, the city's wind circulation plays a significant role in improving the LST of the city [48]. The reasonable shape and structure model of urban buildings can form a favorable environment for wind circulation. Through the atmospheric circulation, the heat island can exchange with the air in the surrounding areas, to reduce the temperature of the city. Therefore, we can effectively arrange the plane and facade of the building, and the volume and height should be organically combined, so that the wind can form a circulation within a certain range. Furthermore, the layout of the living area, office area, commercial area, and industrial area in the city should be reasonably considered to build a multi-functional mixed urban system; particularly, the industrial area should vigorously develop new energy to slow down the heat accumulation.

(2) Building urban green infrastructure systems with stable energy cycles. As an important part of green infrastructure, vegetation can fully absorb the solar radiation [49]. It converts most of the radiant energy into chemical energy in photosynthesis, so as to transfer a large amount of heat energy, which promotes energy conversion and material

circulation of the urban ecosystem [50,51]. More near natural habitats can be created through reasonable green infrastructure layout, increasing ventilation corridors, and reducing unnecessary hard environment to provide good living conditions for people to coexist harmoniously with other organisms.

(3) Developing new energy technologies. A large amount of man-made heat energy is one of the important factors in the rise of LST, such as the use of air-conditioning in summer, industrial exhaust gas, and automobile exhaust emissions. Therefore, it is necessary to reduce the use of nonrenewable energy, promote the use of new energy, and continue to develop new energy such as wind energy, solar energy, and other renewable energy sources for improving the urban habitat [52–54]. Among them, biomass energy has the advantage of being renewable, clean and low-carbon, and with an abundance of raw materials. The total amount of waste generated in Chinese cities every year is abundant. If it is not used rationally, it is easy to waste resources, even pollute the environment, and reduce the quality of urban habitat. Therefore, the development of biomass technology is particularly important. Currently, the bioenergy utilization technologies currently under development in China mainly include thermochemical conversion technology, biochemical conversion technology, biomass briquetting technology, etc. In addition, the use of geothermal technology is very common in some European countries, and we can use geothermal resources to serve construction. Through the development of ground source heat pump technology, the use of air conditioning in buildings can be greatly reduced.

(4) Optimizing the spatial pattern of biodiversity conservation according to local conditions. On the one hand, LST should be adjusted to reduce its negative effects on biological survival. On the other hand, the nature reserves should be optimized, and the protection and supervision of priority areas should be strengthened according to the life history, genes, physiological characteristics, and geographical scope of local organisms with regard to changes in LST [55,56]. For example, we can improve the ex situ conservation system of biodiversity and build ecological corridors that promote species migration and gene exchange, so as to solve the outstanding problems such as urban heat islands, fragmentation and isolated islands of natural habitat, and low ecological connectivity [57,58]. Moreover, sustainable utilization technologies of the biological resources at all levels and types of habitats should be deeply studied [59,60].

6. Conclusions

Based on the analysis of the spatio-temporal differentiation characteristics of habitat quality in Zhengzhou from 2000 to 2015, the critical indicator of LST was introduced to study whether there is a gradient effect and spatial correlation between habitat quality and LST. The main conclusions are as follows:

(1) From 2000 to 2015, the urbanization of Zhengzhou was in a period of accelerated development. The mean value of habitat quality gradually decreased from 0.361 to 0.304, and a large number of middle and middle-low level habitats (II and III) turned into low habitats (I), but the degradation trend slowed down in 2010. Spatially, middle-low level habitats (II) were the most widely distributed, mainly in the low-elevation plains in the east, and low-level habitats (I) were concentrated in the urban built-up areas; middle- and high-level habitats (III, IV, and V) were mainly distributed in the high-altitude western mountainous and hilly regions and northern waters.

(2) LST dropped slightly from 2000 to 2005, rose significantly in 2010, and dropped slightly again in 2015. Spatially high gradients (8, 9, 10) were distributed in the central urban area and the southern airport area, medium gradients (5, 6, 7, 8) were widely distributed, and low gradients (1, 2, 3, 4) were concentrated in the higher elevations of the western mountains and northern waters.

(3) Habitat quality in different periods showed different trends along the LST gradient. High values of habitat quality were concentrated in low gradients (2, 3, 4). Low values were concentrated in high gradients (8, 9, 10), indicating that a slightly lower LST can ensure habitat stability. The excessively high LST may have reduced surface runoff, water

shortage, and other natural restrictive factors, resulting in the decline of the functional structure of the habitat and the ability to support the biological community.

(4) The global Moran's I of the four periods were -0.430, -0.313, -0.323, and -0.311. Overall, there was a significant spatial negative correlation: the higher the LST, the lower the habitat quality. However, at different spatial locations, the distribution and agglomeration characteristics of LST and habitat quality can be divided into the following four types of agglomeration: the high-high type was scattered mainly in the western and northern regions of the study area; the high-low type was mainly distributed in densely populated and actively constructed central urban areas; the low-low type was mainly distributed in the urban-rural intersection and small- and medium-sized rural settlements; and the low-high type was mainly distributed in the western mountain hills and the northern waters and river banks.

(5) Combined with the research results and the existing policies of local authorities, we put forward the following suggestions to improve habitat and biodiversity: carrying out scientific urban planning and building design; building the urban green infrastructure systems with stable energy cycles; developing new energy technologies; and optimizing the spatial pattern of biodiversity conservation. Our innovative research has drawn some laws about the spatial correlation between Zhengzhou's habitat quality and LST, but it is relatively limited. More potential factors will be added in the follow-up research to explore the driving mechanism of urban habitat quality to provide systematic theoretical support for habitat and biodiversity protection in other similar cities.

Author Contributions: Conceptualization, Y.H. and E.X.; methodology, Y.H. and E.X.; software, E.X. and C.L.; validation, Y.H., E.X., G.K. and G.T.; formal analysis, Y.H. and E.X.; investigation, Y.H. and C.L.; resources, Y.H.; data curation, G.K.; writing—original draft preparation, Y.H. and G.K.; writing—review and editing, Y.H., E.X. and G.K.; visualization, C.L. and G.K.; supervision, G.K. and G.T.; project administration, G.T.; funding acquisition, G.T. All authors have read and agreed to the published version of the manuscript.

Funding: This study was supported by Urban–Rural Green Space Resources Control and Landscape Ecological Design Disciplinary Innovation and Talents Introduction Centre Program of Henan, China (GXJD006) and the Key Technology R&D Program of Henan Province (212102310838), the Special Fund for Young Talents in Henan Agricultural University (30500930 and 30501053), the Youth Fund of Ministry of Education Laboratory for Earth Surface Processes of Peking University.

Data Availability Statement: The data presented in this study are available on request from the first author.

Acknowledgments: We would like to thank the International Joint Laboratory of Landscape Architecture for their support and Henan Agricultural University for their great help. We also thank the Institute of Geographical Sciences and Natural Resource Research, Chinese Academy of Sciences, and United States Geological Survey.

Conflicts of Interest: The authors declare no conflict of interest.

References

1. Di Febbraro, M.; Sallustio, L.; Vizzarri, M.; De Rosa, D.; De Lisio, L.; Loy, A.; Eichelberger, B.A.; Marchetti, M. Expert-Based and Correlative Models to Map Habitat Quality: Which Gives Better Support to Conservation Planning? *Glob. Ecol. Conserv.* **2018**, *16*, e00513. [CrossRef]
2. Mori, A.S.; Dee, L.E.; Gonzalez, A.; Ohashi, H.; Isbell, F. Biodiversity–Productivity Relationships Are Key to Nature-Based Climate Solutions. *Nat. Clim. Chang.* **2021**, *11*, 543–550. [CrossRef]
3. Lavergne, S.; Mouquet, N.; Thuiller, W.; Ronce, O. Biodiversity and Climate Change: Integrating Evolutionary and Ecological Responses of Species and Communities. *Annu. Rev. Ecol. Evol. Syst.* **2010**, *41*, 321–350. [CrossRef]
4. Hemanth Kumar, N.K.; Murali, M.; Girish, H.V.; Chandrashekar, S.; Amruthesh, K.N.; Sreenivasa, M.Y.; Jagannath, S. 2—Impact of Climate Change on Biodiversity and Shift in Major Biomes. In *Global Climate Change*; Singh, S., Singh, P., Rangabhashiyam, S., Srivastava, K.K., Eds.; Elsevier: Amsterdam, The Netherlands, 2021; pp. 33–44, ISBN 978-0-12-822928-6.
5. Nath, S.; Shyanti, R.K.; Nath, Y. 4—Influence of Anthropocene Climate Change on Biodiversity Loss in Different Ecosystems. In *Global Climate Change*; Singh, S., Singh, P., Rangabhashiyam, S., Srivastava, K.K., Eds.; Elsevier: Amsterdam, The Netherlands, 2021; pp. 63–78, ISBN 978-0-12-822928-6.

6.   Li, Z.-L.; Tang, B.-H.; Wu, H.; Ren, H.; Yan, G.; Wan, Z.; Trigo, I.F.; Sobrino, J.A. Satellite-Derived Land Surface Temperature: Current Status and Perspectives. *Remote Sens. Environ.* **2013**, *131*, 14–37. [CrossRef]

7.   Gillespie, A.; Rokugawa, S.; Matsunaga, T.; Cothern, J.S.; Hook, S.; Kahle, A.B. A Temperature and Emissivity Separation Algorithm for Advanced Spaceborne Thermal Emission and Reflection Radiometer (ASTER) Images. *IEEE Trans. Geosci. Remote Sens.* **1998**, *36*, 1113–1126. [CrossRef]

8.   Zhou, D.; Xiao, J.; Bonafoni, S.; Berger, C.; Deilami, K.; Zhou, Y.; Frolking, S.; Yao, R.; Qiao, Z.; Sobrino, J.A. Satellite Remote Sensing of Surface Urban Heat Islands: Progress, Challenges, and Perspectives. *Remote Sens.* **2019**, *11*, 48. [CrossRef]

9.   Friedl, M.A. Forward and Inverse Modeling of Land Surface Energy Balance Using Surface Temperature Measurements. *Remote Sens. Environ.* **2002**, *79*, 344–354. [CrossRef]

10.   Qi, Y.; Lian, X.; Wang, H.; Zhang, J.; Yang, R. Dynamic Mechanism between Human Activities and Ecosystem Services: A Case Study of Qinghai Lake Watershed, China. *Ecol. Indic.* **2020**, *117*, 106528. [CrossRef]

11.   Rushing, C.S.; Royle, J.A.; Ziolkowski, D.J.; Pardieck, K.L. Migratory Behavior and Winter Geography Drive Differential Range Shifts of Eastern Birds in Response to Recent Climate Change. *Proc. Natl. Acad. Sci. USA* **2020**, *117*, 12897–12903. [CrossRef] [PubMed]

12.   Johnson, M.D. Measuring Habitat Quality: A Review. *Condor* **2007**, *109*, 489–504. [CrossRef]

13.   Lin, Y.-P.; Lin, W.-C.; Wang, Y.-C.; Lien, W.-Y.; Huang, T.; Hsu, C.-C.; Schmeller, D.S.; Crossman, N.D. Systematically Designating Conservation Areas for Protecting Habitat Quality and Multiple Ecosystem Services. *Environ. Model. Softw.* **2017**, *90*, 126–146. [CrossRef]

14.   Tallis, H.; Lubchenco, J. Working Together: A Call for Inclusive Conservation. *Nature* **2014**, *515*, 27–28. [CrossRef] [PubMed]

15.   Jill, B. Planetary Health: Protecting Nature to Protect Ourselves. Samuel Myers and Howard Frumkin (Eds). *Int. J. Epidemiol.* **2021**, *50*, 697–698. [CrossRef]

16.   Seto, K.C.; Guneralp, B.; Hutyra, L. Global Forecasts of Urban Expansion to 2030 and Direct Impacts on Biodiversity and Carbon Pools. *Proc. Natl. Acad. Sci. USA* **2012**, *109*, 16083–16088. [CrossRef]

17.   Kowarik, I. Novel Urban Ecosystems, Biodiversity, and Conservation. *Environ. Pollut.* **2011**, *159*, 1974–1983. [CrossRef] [PubMed]

18.   Feest, A.; Aldred, T.D.; Jedamzik, K. Biodiversity Quality: A Paradigm for Biodiversity. *Ecol. Indic.* **2010**, *10*, 1077–1082. [CrossRef]

19.   Magurran, A.E.; Baillie, S.R.; Buckland, S.T.; Dick, J.M.; Elston, D.A.; Scott, E.M.; Smith, R.I.; Somerfield, P.J.; Watt, A.D. Long-Term Datasets in Biodiversity Research and Monitoring: Assessing Change in Ecological Communities through Time. *Trends Ecol. Evol.* **2010**, *25*, 574–582. [CrossRef]

20.   Sherrouse, B.C.; Semmens, D.J.; Clement, J.M. An Application of Social Values for Ecosystem Services (SolVES) to Three National Forests in Colorado and Wyoming. *Ecol. Indic.* **2014**, *36*, 68–79. [CrossRef]

21.   Sharps, K.; Masante, D.; Thomas, A.; Jackson, B.; Redhead, J.; May, L.; Prosser, H.; Cosby, B.; Emmett, B.; Jones, L. Comparing Strengths and Weaknesses of Three Ecosystem Services Modelling Tools in a Diverse UK River Catchment. *Sci. Total Environ.* **2017**, *584*, 118–130. [CrossRef] [PubMed]

22.   Huang, Z.; Bai, Y.; Alatalo, J.M.; Yang, Z. Mapping Biodiversity Conservation Priorities for Protected Areas: A Case Study in Xishuangbanna Tropical Area, China. *Biol. Conserv.* **2020**, *249*, 108741. [CrossRef]

23.   Sohn, H.-J.; Kim, D.-H.; Kim, N.-Y.; Hong, J.-P.; Song, Y.-K. Evaluation indicators for the restoration of degraded urban ecosystems and the analysis of restoration performance. *J. Korean Soc. Environ. Restor. Technol.* **2019**, *22*, 97–114. [CrossRef]

24.   Kija, H.K.; Ogutu, J.O.; Mangewa, L.J.; Bukombe, J.; Verones, F.; Graae, B.J.; Kideghesho, J.R.; Said, M.Y.; Nzunda, E.F. Spatio-Temporal Changes in Wildlife Habitat Quality in the Greater Serengeti Ecosystem. *Sustainability* **2020**, *12*, 2440. [CrossRef]

25.   Yu, W.; Ji, R.; Han, X.; Chen, L.; Feng, R.; Wu, J.; Zhang, Y. Evaluation of the Biodiversity Conservation Function in Liaohe Delta Wetland, Northeastern China. *J. Meteorol. Res.* **2020**, *34*, 798–805. [CrossRef]

26.   Benez-Secanho, F.J.; Dwivedi, P. Analyzing the Provision of Ecosystem Services by Conservation Easements and Other Protected and Non-Protected Areas in the Upper Chattahoochee Watershed. *Sci. Total Environ.* **2020**, *717*, 137218. [CrossRef]

27.   Rimal, B.; Sharma, R.; Kunwar, R.; Keshtkar, H.; Stork, N.E.; Rijal, S.; Rahman, S.A.; Baral, H. Effects of Land Use and Land Cover Change on Ecosystem Services in the Koshi River Basin, Eastern Nepal. *Ecosyst. Serv.* **2019**, *38*, 100963. [CrossRef]

28.   Dai, L.; Li, S.; Lewis, B.J.; Wu, J.; Yu, D.; Zhou, W.; Zhou, L.; Wu, S. The Influence of Land Use Change on the Spatial–Temporal Variability of Habitat Quality between 1990 and 2010 in Northeast China. *J. For. Res.* **2019**, *30*, 2227–2236. [CrossRef]

29.   Zhang, T.; Gao, Y.; Li, C.; Xie, Z.; Chang, Y.; Zhang, B. How Human Activity Has Changed the Regional Habitat Quality in an Eco-Economic Zone: Evidence from Poyang Lake Eco-Economic Zone, China. *Int. J. Environ. Res. Public Health* **2020**, *17*, 6253. [CrossRef] [PubMed]

30.   Long, H.; Liu, Y.; Hou, X.; Li, T.; Li, Y. Effects of Land Use Transitions Due to Rapid Urbanization on Ecosystem Services: Implications for Urban Planning in the New Developing Area of China. *Habitat Int.* **2014**, *44*, 536–544. [CrossRef]

31.   Zhu, C.; Zhang, X.; Zhou, M.; He, S.; Gan, M.; Yang, L.; Wang, K. Impacts of Urbanization and Landscape Pattern on Habitat Quality Using OLS and GWR Models in Hangzhou, China. *Ecol. Indic.* **2020**, *117*, 106654. [CrossRef]

32.   Yohannes, H.; Soromessa, T.; Argaw, M.; Dewan, A. Spatio-Temporal Changes in Habitat Quality and Linkage with Landscape Characteristics in the Beressa Watershed, Blue Nile Basin of Ethiopian Highlands. *J. Environ. Manag.* **2021**, *281*, 111885. [CrossRef] [PubMed]

33.   Wang, H.; Tang, L.; Qiu, Q.; Chen, H. Assessing the Impacts of Urban Expansion on Habitat Quality by Combining the Concepts of Land Use, Landscape, and Habitat in Two Urban Agglomerations in China. *Sustainability* **2020**, *12*, 4346. [CrossRef]

34. Song, S.; Liu, Z.; He, C.; Lu, W. Evaluating the Effects of Urban Expansion on Natural Habitat Quality by Coupling Localized Shared Socioeconomic Pathways and the Land Use Scenario Dynamics-Urban Model. *Ecol. Indic.* **2020**, *112*, 106071. [CrossRef]
35. Nematollahi, S.; Fakheran, S.; Kienast, F.; Jafari, A. Application of InVEST Habitat Quality Module in Spatially Vulnerability Assessment of Natural Habitats (Case Study: Chaharmahal and Bakhtiari Province, Iran). *Environ. Monit. Assess.* **2020**, *192*, 487. [CrossRef] [PubMed]
36. Zhang, H.; Zhang, C.; Hu, T.; Zhang, M.; Ren, X.; Hou, L. Exploration of Roadway Factors and Habitat Quality Using InVEST. *Transp. Res. Part D Transp. Environ.* **2020**, *87*, 102551. [CrossRef]
37. Zhang, H.B.; Liu, Y.Q.; Xu, Y.; Han, S.; Wang, J. Impacts of Spartina Alterniflora Expansion on Landscape Pattern and Habitat Quality: A Case Study in Yancheng Coastal Wetland, China. *Appl. Ecol. Environ. Res.* **2020**, *18*, 4669–4683. [CrossRef]
38. Liu, J.; Zhang, Z.; Xu, X.; Kuang, W.; Zhou, W.; Zhang, S.; Li, R.; Yan, C.; Yu, D.; Wu, S.; et al. Spatial Patterns and Driving Forces of Land Use Change in China during the Early 21st Century. *J. Geogr. Sci.* **2010**, *20*, 483–494. [CrossRef]
39. Xu, X.; Liu, J.; Zhang, S.; Li, R.; Yan, C.; Wu, S. China multi period land use and land cover remote sensing monitoring data set (CNLUCC). *Data Regist. Publ. Syst. Resour. Environ. Sci. Data Cent. Acad. Sci.* **2018**. [CrossRef]
40. Nie, C.; Yang, J.; Huang, C. Assessing the Habitat Quality of Aquatic Environments in Urban Beijing. *Procedia Environ. Sci.* **2016**, *36*, 162–168. [CrossRef]
41. Lee, D.J.; Jeon, S.W. Estimating Changes in Habitat Quality through Land-Use Predictions: Case Study of Roe Deer (Capreolus Pygargus Tianschanicus) in Jeju Island. *Sustainability* **2020**, *12*, 123. [CrossRef]
42. Choudhary, A.; Deval, K.; Joshi, P.K. Study of Habitat Quality Assessment Using Geospatial Techniques in Keoladeo National Park, India. *Environ. Sci. Pollut. Res.* **2021**, *28*, 14105–14114. [CrossRef] [PubMed]
43. Li, X.; Yu, X.; Wu, K.; Feng, Z.; Liu, Y.; Li, X. Land-Use Zoning Management to Protecting the Regional Key Ecosystem Services: A Case Study in the City Belt along the Chaobai River, China. *Sci. Total Environ.* **2020**, *24*, 143167. [CrossRef] [PubMed]
44. Tomlinson, C.J.; Chapman, L.; Thornes, J.E.; Baker, C. Remote Sensing Land Surface Temperature for Meteorology and Climatology: A Review. *Meteorol. Appl.* **2011**, *18*, 296–306. [CrossRef]
45. Sobrino, J.A.; Jiménez-Muñoz, J.C.; Paolini, L. Land Surface Temperature Retrieval from LANDSAT TM 5. *Remote Sens. Environ.* **2004**, *90*, 434–440. [CrossRef]
46. Moran, P.A. Notes on Continuous Stochastic Phenomena. *Biometrika* **1950**, *37*, 17–23. [CrossRef] [PubMed]
47. Anselin, L. Local Indicators of Spatial Association—LISA. *Geogr. Anal.* **2010**, *27*, 93–115. [CrossRef]
48. Mills, G. Urban Climatology: History, Status and Prospects. *Urban Clim.* **2014**, *10*, 479–489. [CrossRef]
49. Hua, L.; Zhang, X.; Nie, Q.; Sun, F.; Tang, L. The Impacts of the Expansion of Urban Impervious Surfaces on Urban Heat Islands in a Coastal City in China. *Sustainability* **2020**, *12*, 475. [CrossRef]
50. Sturiale, L.; Scuderi, A. The Role of Green Infrastructures in Urban Planning for Climate Change Adaptation. *Climate* **2019**, *7*, 119. [CrossRef]
51. Kato, M.; Yoshizaki, S.; Hashida, S.; Lee, K.; Suzuki, H. The Evaluation of Tree Species of Urban Greening in Japan for Increasing the Bio-Diversity in Urban Area. *J. Jpn. Soc. Reveg. Technol.* **2016**, *42*, 3–8. [CrossRef]
52. Belaud, J.P.; Adoue, C.; Vialle, C.; Chorro, A.; Sablayrolles, C. A Circular Economy and Industrial Ecology Toolbox for Developing an Eco-Industrial Park: Perspectives from French Policy. *Clean Technol. Environ. Policy* **2019**, *21*, 967–985. [CrossRef]
53. Kardooni, R.; Yusoff, S.B.; Kari, F.B.; Moeenizadeh, L. Public Opinion on Renewable Energy Technologies and Climate Change in Peninsular Malaysia. *Renew. Energy* **2018**, *116*, 659–668. [CrossRef]
54. Qazi, A.; Hussain, F.; Rahim, N.A.B.D.; Hardaker, G.; Alghazzawi, D.; Shaban, K.; Haruna, K. Towards Sustainable Energy: A Systematic Review of Renewable Energy Sources, Technologies, and Public Opinions. *IEEE Access* **2019**, *7*, 63837–63851. [CrossRef]
55. Firake, D.M.; Lytan, D.; Behere, G.T. Bio-Diversity and Seasonal Activity of Arthropod Fauna in Brassicaceous Crop Ecosystems of Meghalaya, North East India. *Mol. Entomol.* **2013**, *3*, 18–22. [CrossRef]
56. Beninde, J.; Veith, M.; Hochkirch, A. Biodiversity in Cities Needs Space: A Meta-Analysis of Factors Determining Intra-Urban Biodiversity Variation. *Ecol. Lett.* **2015**, *18*, 581–592. [CrossRef]
57. Krosby, M.; Tewksbury, J.; Haddad, N.M.; Hoekstra, J. Ecological Connectivity for a Changing Climate. *Conserv. Biol.* **2010**, *24*, 1686–1689. [CrossRef] [PubMed]
58. Nor, A.N.M.; Corstanje, R.; Harris, J.A.; Grafius, D.R.; Siriwardena, G.M. Ecological Connectivity Networks in Rapidly Expanding Cities. *Heliyon* **2017**, *3*, e00325. [CrossRef] [PubMed]
59. Antar, M.; Lyu, D.; Nazari, M.; Shah, A.; Zhou, X.; Smith, D.L. Biomass for a Sustainable Bioeconomy: An Overview of World Biomass Production and Utilization. *Renew. Sustain. Energy Rev.* **2021**, *139*, 110691. [CrossRef]
60. Koukios, E.; Monteleone, M.; Carrondo, M.J.T.; Charalambous, A.; Girio, F.; Hernández, E.L.; Mannelli, S.; Parajó, J.C.; Polycarpou, P.; Zabaniotou, A. Targeting Sustainable Bioeconomy: A New Development Strategy for Southern European Countries. The Manifesto of the European Mezzogiorno. *J. Clean. Prod.* **2018**, *172*, 3931–3941. [CrossRef]

forests

Article

Exploring the Determinants of Urban Green Space Utilization Based on Microblog Check-In Data in Shanghai, China

Dan Chen [1], Xuewen Long [1], Zhigang Li [2], Chuan Liao [3], Changkun Xie [1] and Shengquan Che [1,*]

[1] Department of Landscape Architecture, School of Design, Shanghai Jiao Tong University, Shanghai 200240, China; danchen.gator@sjtu.edu.cn (D.C.); xuewen_long@foxmail.com (X.L.); xiechangkun@sjtu.edu.cn (C.X.)
[2] Department of Urban Planning, School of Architectural and Artistic Design, Henan Polytechnic University, Jiaozuo 454000, China; zhigangli86@hpu.edu.cn
[3] School of Sustainability, College of Global Futures, Arizona State University, Tempe, AZ 85281, USA; cliao29@asu.edu
* Correspondence: chsq@sjtu.edu.cn; Tel.: +86-15618931615

Abstract: Urban green space has significant social, ecological, cultural and economic value. This study uses social media data to examine the spatiotemporal utilization of major parks in Shanghai and explore the determinants of their recreational attraction. Methods: Based on microblog check-in data between 2012 and 2018 across 17 parks in Shanghai, we investigated the patterns at different temporal scales (weekly, seasonal and annual) and across workdays and weekends by using log-linear regression models. Results: Our findings indicate that both internal and external factors affect park utilization. In particular, the presence of sports facilities significantly contributes to higher visit frequency. Factors such as the number of subway stations nearby, scenic quality and popularity have a positive impact on check-in numbers, while negative factors affecting park use are number of roads, ticket price and average surrounding housing price. Across different temporal scales, the use patterns of visitors have obvious seasonal and monthly tendencies, and the differences of workday and weekend models lie in external factors' impacts. Conclusions: In order to achieve the goal of better serving the visitors, renewal of urban green spaces in megacities should consider these influential factors, increase sports facilities, subway stations nearby and improve scenic quality, popularity and water quality. This study on spatiotemporal utilization of urban parks can help enhance comprehensive functions of urban parks and be helpful for urban renewal strategies.

Keywords: social media data; urban green space; spatiotemporal utilization; Shanghai

Citation: Chen, D.; Long, X.; Li, Z.; Liao, C.; Xie, C.; Che, S. Exploring the Determinants of Urban Green Space Utilization Based on Microblog Check-In Data in Shanghai, China. *Forests* **2021**, *12*, 1783. https://doi.org/10.3390/f12121783

Academic Editors: Manuel Esperon-Rodriguez and Tina Harrison

Received: 13 November 2021
Accepted: 14 December 2021
Published: 16 December 2021

Publisher's Note: MDPI stays neutral with regard to jurisdictional claims in published maps and institutional affiliations.

1. Introduction

Urban green space plays a crucial role in urban residents' leisure and recreational activities [1]. Services provided by urban green space contribute to both physical and mental health of park users [2–4], enhance the value of surrounding real property [5] and mitigate the effect of urban heat islands [6,7]. For megacities such as Shanghai, there are few opportunities to create new urban green space in its core urban area due to high building density. In order to maximize the benefits of urban green space, it is crucial for urban planners and decision makers to improve the existing parks to better serve citizens and encourage park utilization. To achieve this goal, it is necessary to understand the current situation of park utilization and its determinants.

Previous studies have taken both qualitative and quantitative approaches to investigate urban green space utilization. Qualitative studies such as interviews, observations and records often involve purposeful sampling of participants and settings [2] and convert the interpretations from language and actions to data and conclusions. Surveys are the most popular approach used in quantitative studies to acquire park use data [8,9]. A combination of behavior mapping and GIS (Geographic Information System) techniques can be used to help generate park use patterns by using spatial analysis and visualization [10].

These methods can be used for in-depth evaluation of recreation space, recreation type and behavior in urban parks. Senetra et al. (2018) analyzed spatial distribution, influence and quality of urban green space, finding that green spaces were unevenly distributed and not all residents have equal access to high-quality public greens [11]. However, it is difficult to conduct comprehensive evaluation of multiple urban parks at a broader scale due to intensive labor requirements and the short time span of the survey data.

Based on findings from previous studies, major factors that influence urban green space utilization include both internal and external characteristics of the parks. The internal characteristics of urban parks reflect the quality of parks, which is related to the planning and design, construction quality and maintenance management of parks. A review study by McCormack et al. (2010) indicates that safety, aesthetics, amenities and maintenance are important for encouraging park use. Several studies found that well-designed public open space can promote physical activity in the community and may contribute to the health of local residents [10,12]. Research has shown that the quality of paths in parks, the allocation of infrastructure, the proportion of water in green space and the surrounding environment, among many other factors, have impacts on the population's physical activity. Meanwhile, aesthetic value, safety and other indicators also have positive impacts [13,14]. Besenyi et al.'s research has shown that there is a positive correlation between the environmental elements of urban parks and residents' outdoor physical exercise; that is, parks and green spaces that are high quality of larger scales have a positive impact on people's physical activity [8].

The external factors of a park are mainly related to the location, traffic convenience, economic development status and population density as well as land use types in the surrounding areas. Accessibility of parks is a very influential determinant [2]. Besenyi et al., based on an online survey, concluded that higher accessibility would encourage people to engage in physical activity [8]. Research has shown perceptions of park availability and usage by friends are associated with greater park use [8]. Few studies have reported on the impact of external factors of parks use in addition to accessibility; however, subjective outdoor thermal comfort has been found to impact urban park utilization in Hong Kong [15].

Over the past decade, a considerable amount of data on peoples' location and movement have been collected due to mobile communication devices, position perception technologies and social media. Social media data are widely used, including that from platforms such as Twitter [16,17], Foursquare, Sina microblog [18–20], Flickr [21] and Panoramio [22]. Compared with data obtained through surveys, digital data include the advantage of having a larger amount of information that is highly accurate and richer. This provides opportunities for innovation in urban research. Existing studies have identified the practical significance of using social media data analysis in urban residents' activities. Social media data can be used as a valuable data source for urban space research and analysis of residents' activities and, thus, address shortcomings such as small sample size and short time span in survey research.

Recent research conducted spatiotemporal analysis on urban park recreation verified the feasibility of using Big Data. Most relevant applications of social media data mainly explore the spatial pattern of urban residents' gathering, dispersion and flow relating with time, space, activities and several other aspects [23,24]. The applications of social media data in urban areas are mainly based on GIS platforms, using cluster analysis, kernel density analysis and other spatial analysis and calculation methods to identify the spatial structure and characteristics [25]. The hourly real-time Tencent user density (RTUD) data from social media are used to analyze the spatiotemporal distribution of urban park users in the study by Chen et al., and the factors that may be associated with the user density in urban parks were studied by a group of linear regression models in Shenzhen, China [5]. Roberts proposed a method of crowdsourcing in urban green space using Twitter data and identified the events in urban green space that bring cultural benefits, such as individual and community identity, as well as cultural activities participation [26]. In Spain, popular public spaces in the urban fabric of the city of Murcia were studied, with data sources from Foursquare and Twitter [27]. Roberts et al. used Twitter data to assess the variance of physical activity engagement between summer and winter seasons [28]. Another study analyzed mobile phone data derived from 10 million daily active users to explore spatiotemporal activity patterns of users in Central Park, New York, USA, and found that regions with established amenities and points of interest demonstrate a higher record of shared experience [29].

However, most previous social media studies focused on spatiotemporal analysis and city-scale issues rather than exploring the determinants of urban parks utilization. Using social media data, this paper explores the determinants of urban park utilization. Based on our findings, we propose optimization countermeasures, suggestions for the planning and design of the parks in terms of each determinant and the advantages and shortcomings of social media data in research of urban public space service functions.

The purpose of this study is to take advantage of social media data to indicate the internal and external characteristic influencing factors of urban parks on park utilization. Based on the interactive relationship between recreation space and environmental behavior, influencing factors of the use of public green space in a high-density metropolis can be identified. This puts forward the corresponding urban renewal optimization countermeasures, which also has reference significance for planning and design of urban green space in other megacities.

2. Materials and Methods

2.1. Study Area

In 2020, Shanghai had a total population of 24 million, and the per capita area of park green space was 8.5 square meters. Urban public green space is a scarce commodity in the high-density urban center of Shanghai. There are at least 2–5 comprehensive parks in each administration district, which mainly serve the residents of the district and attract visitors from other districts as well. These parks generally cover an area of more than 5 hectares and have a large number of visitors. They have very important social, ecological and recreational services. Previous studies related to urban green spaces focused on spatial layout and accessibility [8]. However, the urban layout of Shanghai has been basically stable and green space is in the stage of slow growth. We should review green spaces from the perspective of urban renewal and micro design. Social media data provide a wide range and high amount of data for the study of public space in a high-density metropolis such as Shanghai and help to scientifically analyze the influencing factors of urban park green space use.

In this research, we studied the central area of Shanghai, China. According to the "Master Urban Planning of Shanghai," the central area of Shanghai is within the outer ring road, with an area of about 660 square kilometers (as shown in Figure 1, including Zone 1 to Zone 8, and Table 1).

Using the "Regulations on Administration of Parks in Shanghai," "Guidance on the Implementation of Classification and Grading of Urban Parks in Shanghai (trial)" and "Classification Standards of Urban Green Space (by central government)," we assessed the urban parks in the central area of Shanghai from 3 aspects, namely park type, park area and park influence. For park type, we chose comprehensive parks rather than rural parks; for park area, most of parks we chose had an area large than 5 ha; and for park influence, parks with diverse facilities for recreation and with a certain number of user in terms of microblog check-in data were used. We then identified 17 comprehensive parks as our research subjects (Figure 2).

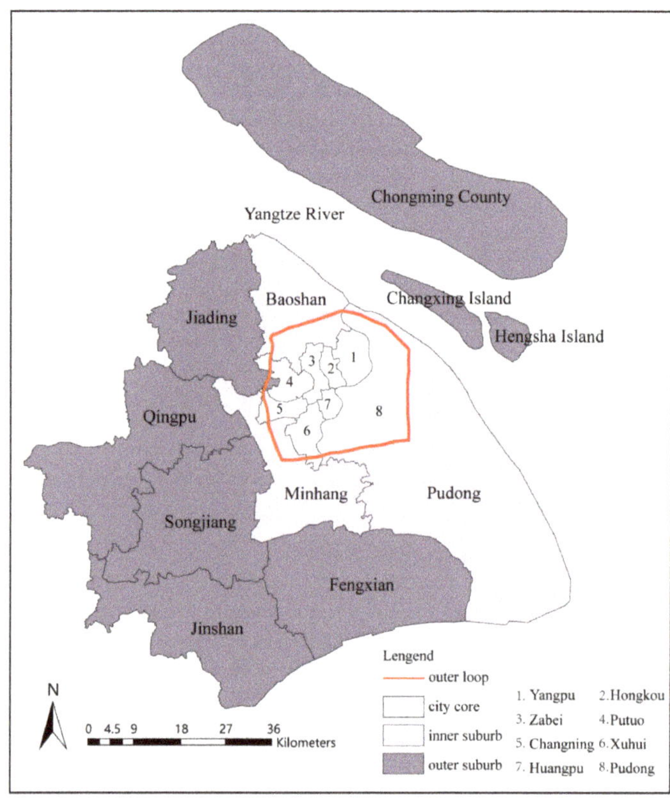

Figure 1. Scope of Central Shanghai.

Table 1. List of 17 comprehensive parks in downtown Shanghai.

No.	District	Park Name	Opening Time	Area (ha)	Park Ranking	Number of Sports Venues	Popularity	Scenic Quality	Water Area (ha)	Green Coverage Ratio (%)
1	Changning	Zhongshan Park	1914	21.42	4	1	378	4.4	1.22	91
2		Tian Shan Park	1959	6.80	4	1	-	4.3	1.67	75
3	Hongkou	Luxun's Park	1896	28.63	4	2	332	4.4	3.00	85
4		Peace Park	1958	16.34	3	1	-	4.5	1.80	85
5	Huangpu	People's Park	1952	10.00	4	1	495	4.4	0.08	90
6		Fuxing Park	1909	8.89	4	0	217	4.5	0.30	88
7		Jing'an Park	1955	3.36	5	0	158	4.4	0.09	90
8	Jing'an	Daning Lingshi Park	2002	68.00	3	2	323	4.6	7.08	85
9		Zhabei Park	1946	6.67	4	1	166	4.3	0.80	85
10		Century Park	1997	140.30	5	2	845	4.7	27.00	80
11	Pudong	Expo Park	2010	23.00	4	1	196	4.4	0.70	70
12		Lujiazui Central Park	1997	10.00	4	1	-	4.6	0.45	90
13	Putuo	Changfeng Park	1959	36.40	5	1	599	4.5	14.3	55
14	Xuhui	Xujiahui Park	2001	8.66	5	1	185	4.4	0.4	90
15		Guilin Park	1931	3.55	4	0	200	4.5	0.08	90
16	Yangpu	Yangpu Park	1957	22.00	4	1	-	4.3	1.9	85
17		Huangxing Park	2000	39.86	3	2	241	4.4	8	75

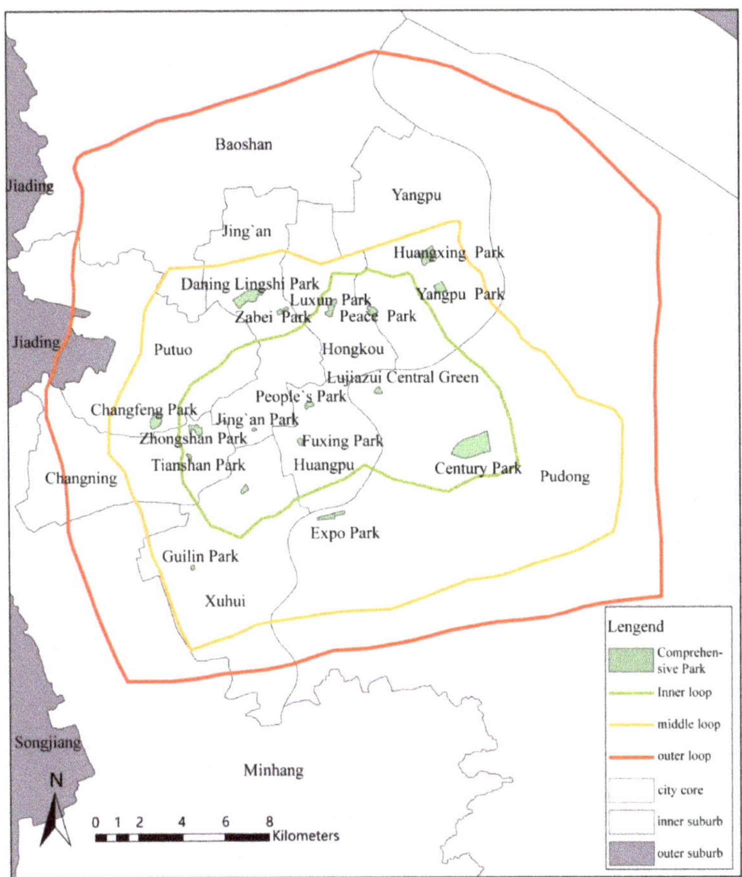

Figure 2. Distribution of 17 comprehensive parks in central Shanghai.

2.2. Selection of Dependent Variable

Sina microblog is one of the social network platforms with the largest number of users in China. According to the 2017 microblog user development report released by the microblog data center, the number of monthly active users on the microblog totaled at 376 million as of September 2017 (https://data.weibo.com/report/reportDetail?id=404, last accessed date (10 December 2021)). The check-ins in the microblog are used as dependent variable and completely records the relevant contents, including geographic information (latitude and longitude coordinates), time, texts and sex of microblog users, reflecting part of the spatiotemporal behavior of users in the city. From the texts of microblog users, it can be found that most of them visited parks for recreation, socialization and relaxation.

2.3. Data Collection

This paper uses Sina microblog check-in data (from the time period 1 January 2012 to 30 June 2018), obtained through the Sina microblog API (Application Programming Interface Application Data Interface); Shanghai downtown comprehensive park data and house price transaction data used in this study were obtained by using web crawler; Shanghai administrative boundary and Shanghai central comprehensive parks data were from the official website of Shanghai Greening and Urban Management Bureau. Parks are classified into two-star, three-star, four-star and five-star by Shanghai Greening and Appearance Bureau, which is the meaning of Park Ranking in Table 1. Number of Sports

Venues was counted as number of sports and fitness venues in each park, including trails, basketball courts, tennis courts, football courts and other sports venues (1 point per item). Popularity is the average number of internet users searching park names by using Baidu over a certain period of time. Scores of Scenic Quality given for the landscape effect of the comprehensive park can be divided into 5 levels: extremely poor (1 point), poor (2 points), general (3 points), good (4 points) and great (5 points). Water area (ha) means the area of water bodies in each park. Green Coverage Ratio (%) means the vertical projection area of vegetation in each park.

Quantification of Variables

A total of 13 independent variables were selected to quantify the check-ins. Summary statistics of the variables are included in Table 2.

Table 2. Summary statistic of variables used.

Variable	Variable Name in Models	Quantification of Variables	Mean (Std. Dev.)	Min. (Max.)
Number of check-ins	Count	Sina microblog check-ins	157.23 (262.29)	0 (4327)
Green coverage ratio	Green	The vertical projection area of vegetation in each park. (%)	82.23 (9.011)	55 (90)
Ticket price	Ticket fee	Ticket price (yuan)	0.71 (2.37)	0 (10)
Area of park	Area	Area of each park. (ha)	26.54 (32.70)	3.36 (140.3)
Area of water	Water	Area of water bodies in each park. (ha)	4.05 (6.823)	0.08 (27)
Popularity (Baidu Index)	Baidu	Over a certain period of time, the average number of internet users searched park names through Baidu.	333.63 (196.24)	158 (845)
Number of subway stations	400 Station	Number of subway stations in the 400-metre buffer zone around the park.	1.06 (0.540)	0 (2)
Number of bus stations	Bus	Number of bus stops in the 400-metre buffer zone around the park.	3.18 (1.724)	0 (6)
Average housing prices	Average HP400	Average housing prices in the 400-metre buffer zone around the park. (Yuan/m^2)	78,627.31 (12,563.05)	62,416 (103,308)
Area of commercial land	ComArea	The area occupied by commercial office in the 400-m buffer zone around the park. (ha)	20.94 (17.74)	4 (72)
Scenic quality	Beauty	Scores given for the landscape effect of the comprehensive park can be divided into 5 levels: extremely poor (1 point), poor (2 points), general (3 points), good (4 points) and great (5 points).	4.447 (0.11)	4.3 (4.7)
Park ranking	Star	Parks are classified into two-star, three-star, four-star and five-star by Shanghai Greening and Appearance Bureau.	4.06 (0.64)	3 (5)
Number of sports venues	Sports Number	The number of sports and fitness venues in each park, including trails, basketball courts, tennis courts, football courts and other sports venues (1 point per item).	1.06 (0.64)	0 (2)
Number of roads	Road	The number of roads in the 400-metre buffer zone around each park.	16 (6.78)	5 (34)

2.4. Research Design

In this study, we collected 214,068 Sina microblog check-in points in 17 comprehensive parks of central Shanghai from 2012 to 2018, among which 210,993 of points are valid (removing those with missing values). Workday and weekend data are separately analyzed to determine differences of park utilization.

The spatiotemporal distribution of microblog checked-ins were analyzed at various temporal scales, such as annual, monthly and intra-day scales, to help identify potential determinants of park use. The spatial and vector data in 17 studied comprehensive parks, the geographic coordinate information in the Sina microblog check-ins and the data on road density, public transportation number and housing price around the parks were input into GIS 10.6 and analyzed by using spatial analysis methods.

In this study, Sina microblog check-ins in comprehensive parks were used as the quantitative indicator of park use, while green coverage ratios, ticket price, area of park, area of water, popularity (Baidu Index), number of bus stations, average housing prices, area of commercial lands, scenic quality, parking ranking, number of sports venues and number of roads were used as the independent variables for analysis with multivariate regression models.

The urban parks utilization data were classified into weekday use data and weekend use data, and the check-ins were divided into yearly, seasonal and monthly groups, which served as the dependent variables of the subsequent stepwise regression model.

2.5. Model Selection

In this study, four linear regression models were considered, which included linear form, logarithmic form, logarithmic linear form and semi-logarithmic form, with the selected variables to test which is the most suitable model. Since some independent variables were coded to have a value of 0 to 1, only linear form and log-linear form were tested. Table 3 indicates that the log-linear model has the highest degree of fitting and the strongest explanatory power. Therefore, log-linear regression was chosen to carry out the multivariate linear regression analysis of comprehensive parks utilization.

Table 3. Summary of four functional forms of linear regression models.

Model Form			Without Dummy Variable		With Dummy Variable	
			Linear	Log-Linear	Linear	Log-Linear
Yearly model	Total	Adjusted R^2	0.29	0.47	0.35	0.56
		Residuals	−117.86	0.12	−117.86	−0.05
		Average	1805.80	6.85	1805.80	6.85
	Workday	Adjusted R^2	0.36	0.49	0.47	0.60
		Residuals	−46.67	−0.22	−47.14	−0.22
		Average	748.32	5.92	1805.80	6.85
	Weekend	Adjusted R^2	0.22	0.44	0.30	0.44
		Residuals	−65.57	0.65	−65.57	0.65
		Average	1057.48	6.32	1805.80	6.85
Seasonal model	Total	Adjusted R^2	0.25	0.65	0.34	0.65
		Residuals	0.00	0.00	−23.83	0.18
		Average	478.92	5.43	478.92	5.43
	Workday	Adjusted R^2	0.19	0.64	0.29	0.71
		Residuals	−4.07	0.00	−4.09	0.00
		Average	281.92	4.90	281.92	4.90
	Weekend	Adjusted R^2	0.28	0.62	0.38	0.72
		Residuals	−11.63	0.00	−10.08	0.00
		Average	196.32	4.47	196.32	4.47

Table 3. *Cont.*

Model Form			Without Dummy Variable		With Dummy Variable	
			Linear	**Log-Linear**	**Linear**	**Log-Linear**
Monthly model	Total	Adjusted R^2	0.23	0.60	0.32	0.70
		Residuals	0.00	0.00	0.00	0.00
		Average	157.23	4.24	157.23	4.24
	Workday	Adjusted R^2	0.16	0.56	0.21	0.67
		Residuals	0.00	0.00	−1.51	0.00
		Average	92.76	3.72	92.76	3.72
	Weekend	Adjusted R^2	0.24	0.55	0.32	0.68
		Residuals	0.00	0.00	0.00	0.00
		Average	64.35	3.31	64.35	3.31

Dummy Variables

From descriptive spatiotemporal distribution of visitors' analysis, we found that year, season and month contributed to the total check-ins of visitors to comprehensive parks. Therefore, dummy variables were created to identify the influences by yearly, seasonal and monthly factors.

3. Results

Table 4 shows yearly, seasonal and monthly regression results of comprehensive park utilization models using Microblog check-ins from 2012 to 2018.

Table 4. Regression results of comprehensive park utilization models.

Model	Yearly				Seasonal				Monthly			
	Non-Std. Coeff.	Standard Coeff.	t	Sig.	Non-Std. Coeff.	Standard Coeff.	t	Sig.	Non-Std. Coeff.	Standard Coeff.	t	Sig.
	B	β			B	β			B	β		
(constant)	1.982	-	2.420	0.018	−39.368	-	−7.768	0.000	−85.494	-	−22.027	0.000
SportsNumber	1.240	0.690	6.244	0.000	1.394	0.727	14.746	0.000	4.861	2.528	23.756	0.000
AverageHP400	0.000048	0.476	4.383	0.000	0.000092	0.754	12.729	0.000	−0.000027	−0.254	−3.446	0.001
Park Ranking	-	-	-	-	-	-	-	-	5.334	2.495	17.294	0.000
Road	−0.062	−0.359	−3.964	0.000	−0.077	−0.418	−9.858	0.000	−0.257	−1.376	−21.567	0.000
2015	−0.774	−0.220	−2.844	0.006	−0.924	−0.234	−6.872	0.000	−1.037	−0.262	−13.446	0.000
2014	−0.755	−0.215	−2.774	0.007	−0.788	−0.200	−5.863	0.000	−0.936	−0.238	−12.184	0.000
ComArea	0.023	0.298	3.429	0.001	0.051	0.626	11.013	0.000	0.011	0.129	2.784	0.005
Bus	-	-	-	-	−0.491	−0.631	−8.521	0.000	0.419	0.533	6.439	0.000
400Station	0.691	0.324	3.914	0.000	0.914	0.402	10.648	0.000	4.483	1.938	21.321	0.000
2016	-	-	-	-	−0.551	−0.140	−4.098	0.000	−0.552	−0.140	−7.185	0.000
October	-	-	-	-					0.476	0.086	4.183	0.000
Beauty	-	-	-	-	8.628	0.622	7.757	0.000	14.110	1.009	20.076	0.000
Ticketfee	-	-	-	-	−0.405	−0.774	−8.717	0.000	−0.324	−0.617	−8.409	0.000
Water	−0.035	−0.203	−2.091	0.040	-	-	-	-	−0.728	−3.919	−17.506	0.000
Baidu Index	-	-	-	-	-	-	-	-	0.012	1.595	15.914	0.000
Spring	-	-	-	-	-	-	-	-	0.269	0.082	4.025	0.000
2012	-	-	-	-	-	-	-	-	−0.277	−0.070	−3.582	0.000
Autumn	-	-	-	-	-	-	-	-	0.227	0.065	2.861	0.004
January	-	-	-	-	-	-	-	-	−0.305	−0.079	−3.221	0.001
2018	-				-				0.279	0.055	2.902	0.004
February	-	-	-	-	-	-	-	-	−0.214	−0.041	−2.198	0.028
Winter	-	-	-	-	−0.427	−0.122	−3.560	0.000	-	-	-	-
Summer	-	-	-	-	−0.343	−0.104	−3.070	0.002	-	-	-	-

3.1. Yearly Comprehensive Park Utilization Model Results

A total of 102 independent variables were calculated in the yearly comprehensive park utilization model. The correlation coefficient R is 0.777, the judgment coefficient R^2 is 0.604 and the adjusted R^2 is 0.558.

According to Table 4, the most influential determinant is the number of sports venues, followed by average housing price, number of roads, area of commercial lands, number of subway stations and area of water. The regression results shows that yearly average check-ins will increase 124% for every increase in the number of sports venues; with every CNY 1000 increase in average house price around the parks, park utilization will increase 4.8%. For each hectare increase in commercial land area, the check-ins will increase 2.3%. For every one station increase in the number of subway stations, park use will increase by 69%. Regarding negative impacts, one more road around the park will contribute to a 6.2% decrease in check-ins, and a per hectare water area increase will result in a 3.5% decrease in park use.

In the yearly workday and weekend models, variables such as the number of sports venues, average housing price and area of commercial land had positive influences on check-ins, while the number of roads had negative impacts.

3.2. Seasonal Comprehensive Park Utilization Model Results

For the seasonal comprehensive park usage model, correlation coefficient R was 0.818, the judgment coefficient R^2 was 0.669 and the adjusted R^2 was 0.656.

According to Table 4, seasonal check-ins will increase by 139% for every increase in the number of sports venues. With every CNY 1000 increase in average house price around the park, check-ins will increase by 9.2%, and for each hectare increase in commercial property, park use will increase by 5.1%. With every one increase in the number of metro stations, check-ins will increase by 91%. A 0.01 increase in scenic quality was associated with an 8.62% increase in check-ins. In contrast, some studied variables had negative impacts: One more road in the area around the park will cause a 7.7% decrease in check-in numbers. Ticket price also has a negative influence on park use, with a 40.5% decrease per CNY increase, and every bus station contributes to a 49% decrease in check-ins. Furthermore, seasonality is a factor in check-ins, with summer and winter having fewer negative effects on check-ins.

For seasonal workday and weekend models, determinants such as number of sports venues, park ranking, Baidu Index, scenic quality and number of subway stations all have positive influences on park use, while the number of roads and green coverage ratio induce negative impacts. In addition, spring and autumn are positive for check-ins in the workday model, and summer and winter seasons are negative for park use in the weekend model. Average housing price is positive in the workday model, and area of commercial lands is positive in the weekend model.

3.3. Monthly Comprehensive Park Utilization Model Results

After removing outliers, a total of 1320 observations were used in the monthly comprehensive park utilization models. The correlation coefficient R was 0.840, the judgment coefficient R^2 was 0.705 and the adjusted R^2 was 0.699.

The regression results show that monthly microblog check-ins will increase 486% for every increase in the number of sports venues. With each hectare increase in commercial land area, park utilization will increase by 1.1%, and every one increase in the number of Metro stations, park use will increase by 448%. The check-ins with a 0.01 increase in scenic quality will increase by 14.11% more and a value of 0.01 increases in park ranking will contribute to 5.33% more check-ins. The check-ins will increase by 1.2% with each increase in Baidu Index; every bus station brings a 41.9% decrease in park utilization. For negative impacts, one additional road around the park will cause a 25.7% decrease in check-ins, and each hectare increase in water area will result in a 72.8% decrease in park use. Ticket price also has a negative influence on check-ins with a 32.4% decrease for CNY 1 increase. Lastly for every CNY 1000 increase in average house price around the parks, the check-ins will decrease by 2.7%.

Monthly workday and weekend models indicate that variables such as the number of sports venues, park ranking, scenic quality, Baidu Index, number of bus stations, number of subway stations and month have positive influences on park use, while ticket price, area of water and number of roads have negative impacts. In addition, spring and autumn were found to be correlated with park use in the workday model, while summer and winter seasons were correlated with use in the weekend model. Average housing price had a positive correlation in the workday model, while the green coverage ratio had a negative correlation. Area of commercial lands had a positive correlation in the weekend model.

3.4. Daily Comprehensive Park Utilization Model Results

According to Figure 3, the time of 1:00–6:00 a.m. is the sleeping time. Numbers of visitors to comprehensive parks increases from 7:00 a.m. to 23:00 p.m. and peaks around 14:00–15:00 p.m., followed by another small check-in peak from 19:00 to 21:00 p.m.

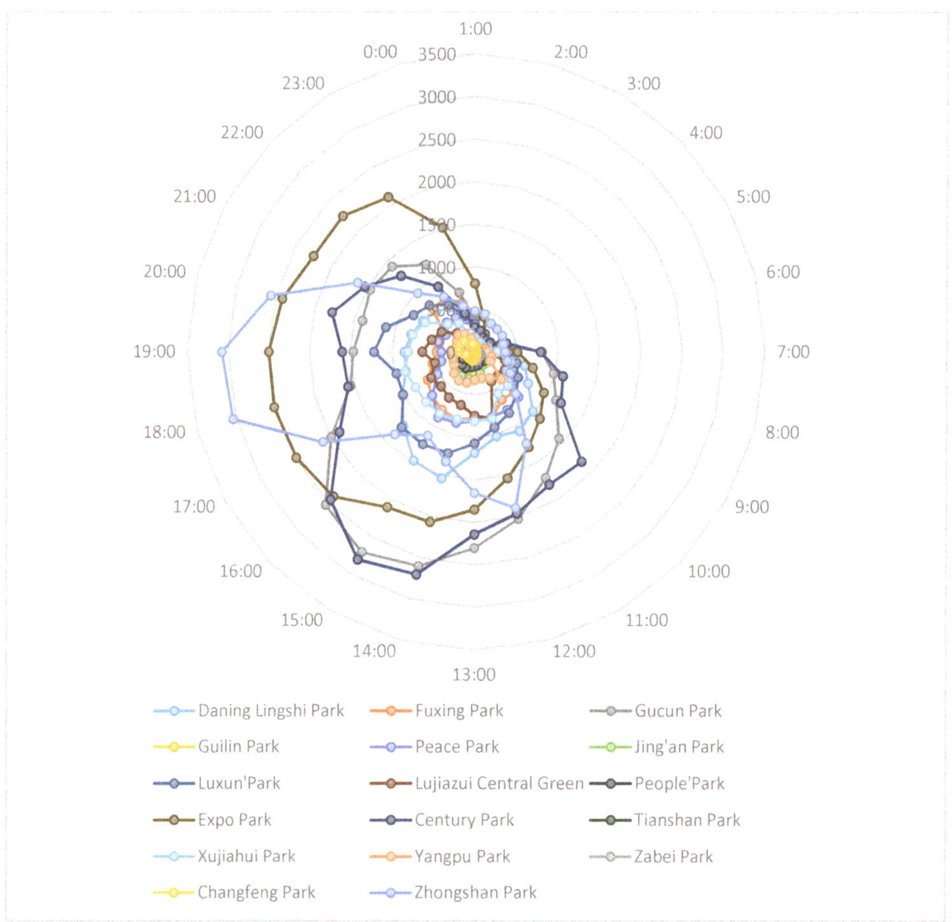

Figure 3. Intraday distributions of check-in data.

The comparison of check-in time between weekdays and weekends is shown in Figure 4. The graphs indicate that recreational time for visitors was generally between 9:00 and 21:00. The check-in peak on weekdays appeared at 19:00, while the peak on weekends was around 15:00, and the number of average daily visitors on weekends significantly increased compared with that on weekdays.

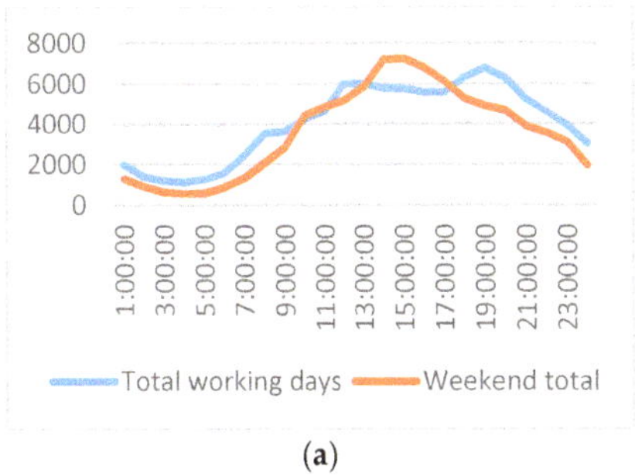

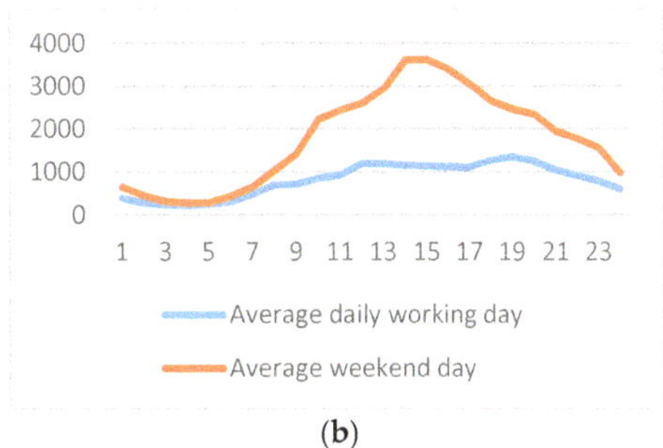

(a)   (b)

Figure 4. Cumulative daily check-ins in comprehensive parks from 2012 to 2018. (**a**) Cummulative check-ins; (**b**) Average daily check-ins.

4. Discussion

Both internal and external factors of urban parks play significant roles in determining park utilization. The number of sports venues is the most influential determinant, followed by the number of subway stations, scenic quality, park ranking, Baidu Index, number of bus stations and area of commercial lands. Factors with the most negative influence on park check-ins are number of roads, area of water, ticket price and average housing price. Seasons, such as spring and autumn, had a positive influence on park use, while check-ins during January and February decreased.

Number of sport venues is the most influential element and had a positive impact on check-ins, while number of roads adjacent to the park had the most negative impact. The demand for sports venues is in line with the health motivation of visitors. The more sports venues there are, the more attractive urban green space will be to visitors.

In general, the number of subway stations, the area of commercial zones and the average housing price were associated with more urban parks utilization. The number of subway stations makes parks more accessible. A larger area of commercial lands around the park creates more attractions for visitors' recreation. This type of mixed land use has been used more for adapted land use planning in recent years and is viewed as a healthier and dynamic land use pattern [30]. At present, conventional commercial facilities, such as restaurants, teahouses, bookstores and snack shops, in parks are insufficient to meet the needs of park visitors, and the commercial office space around the park can make up for this defect. A large number of studies have shown that urban parks have a value-added effect on housing prices [31]. Housing close to urban parks with high access to green space and scenery is more expensive. Dunse et al. (2007) quantified the economic benefits of open space on households in Aberdeen, Scotland, and confirmed the positive influence of public parks on housing price [32]. Therefore, urban parks attract urban residents to gather while simultaneously promoting the value of the surrounding property.

The size of the park is not statistically significant, and the area of water is not always significant; if it is, its impact is negative. Previous studies believe that visitors prefer an urban green space with water body to spaces without any water elements [13,33]. Data from this study indicate that it having more water does not result in more utilization. There are two probable reasons: (1) more water means less space for other use or a large area of water separating the use of the park; (2) water quality is another consideration. The decline in water quality in urban parks in Shanghai [34] may result in less utilization, which is consistent with the study in London [35]. There is a correlation between a park's area and water area and check-ins and, more specifically, in the internal diversity of sites, infrastructure diversity and spatial diversity [11,36].

For seasonal and monthly models, additional variables had influences. Scenic quality always plays an important positive role; park ranking and Baidu Index have positive functions as well, while ticket price and green coverage ratio have a negative effect on attendance. The variable "scenic quality" properly reflects the quantitative relationship with recreational attraction; it had a positive effect on park use. Previous research indicates that landscape naturalness, landscape characteristics, plant collocation and plant richness are all related to aesthetic degree [37]. Therefore, the aesthetic design of the landscape should pay more attention to the visual attraction of the site in order to optimize visitor satisfaction.

Park rankings are related to management quality. The results of Gui (2016) proposed that Shanghai residents place more value on the number of recreational spaces and facilities, the degree of maintenance of the facilities, environmental quality and the functional complexity of recreational space [38]. Urban residents have certain requirements for the management quality of parks, and park rankings help to summarize this management quality. The cost of park tickets is an obstacle to visitors' recreation. Zeng (2015) studied visitor data released by the Shanghai Municipal Administration of Greening and Appearance, and the findings suggest that the rapid growth of the number of visitors after parks offered free admission increased the difficulty of park management. Thus, there is still a balance to be considered on park admission fees.

For seasonal and monthly models, spring and autumn exerted positive influences on the workday model, while summer and winter caused negative influences. This result is consistent with simple descriptive spatiotemporal distribution analysis; that is, visitors prefer to attend parks in the spring and autumn seasons when the weather is nice. For the monthly models, October was found to have a significant positive influence, which is also reasonable because of the National Holiday in China and typically nice weather. The weather in spring and autumn is suitable for recreation, and autumn coincides with the National Day holiday, resulting in higher check-ins; therefore, outstanding seasonal features and improvements to summer and winter landscapes in urban parks may help balance park utilization.

In terms of the seasonal and monthly workday models, spring, autumn seasons and October showed positive influences on check-ins. According to the seasonal and monthly weekends models, the area of commercial land is significant and positive; summer and winter exert negative influences; and for the monthly weekend model, March, October and December are positive. On workdays, nice weather has a significant positive effect on park visitation. The number of bus stations is also important. During weekends, bad weather has significant negative effects, and the area of commercial lands is important. Therefore, we conclude that the differences of workday and weekend models lie in external factors, and the surrounding commercial conditions and good weather are important for park weekend utilization.

Regarding monthly characteristics of visitors' use in comprehensive parks, the peak is in October, which may be related to the impact of collective travel of urban residents during the National Day Golden Week. The check-ins are typically the highest in autumn, leading us to the suggestion that autumn has suitable weather and coincides with the National Day holiday, when people have more leisure time to spend in parks. The climate conditions during spring and autumn encourage park use, while winter and summer have fewer ideal conditions and limit visitors.

Daily analysis provides some interesting findings regarding visitors' behaviors. More people chose to visit the park in the afternoon, and some people visited parks after dinner (Figure 4). Check-ins at 12:00 and 18:00 formed two troughs, indicating that parks have fewer visitors during lunch and dinner hours. We also find that visiting time tendencies are predictable. Restricted by work, the leisure time of visitors is constrained. Overall, park recreation time is concentrated on weekends and golden week holidays. The average number of visitors on weekends is nearly twice compared with that of weekdays. The peak of park use on weekdays is at 19:00 after dinner, and the peak of recreation on weekends is around 15:00 p.m. On weekends, visitors have a higher demand for the recreational landscape environments.

An interesting finding is that the intraday check-in curve of Shanghai Expo Park is significantly different from that of other parks (Figure 3). Check-ins to the park were mainly distributed between 16:00 and 23:00 p.m. because Shanghai Expo Park often holds outdoor activities, such as the Strawberry Music Festival, Jazz Festival and Shanghai Magic Lantern Carnival, which attracts a large number of people from inside and outside Shanghai. As a result, the park had a low local visitor rate.

Social media use is growing fast with technological innovation and subsequent data collection. These large, efficient and growing datasets bring massive information to researchers, which help to inspire research questions and methods. Spatiotemporal datasets from social media allow researchers to study travel behaviors, assess location characteristics, identify tourist spots, assess attractiveness and reveal land use patterns via approaches that were impossible with traditional methods of data collection [39]. In this study, social media data, such as Microblog check-ins, provide researchers a large volume data to analyze yearly, seasonal, monthly and daily utilization situations for several parks in Shanghai, which is difficult to fulfill by using traditional questionnaires and observation methods. It is believed that social media data benefit spatiotemporal studies, especially when the studies include more than one location.

However, it is important to keep in mind the shortcomings of social media. First, the location information may not be exact, resulting in incorrect locational check-ins, especially for the check-ins at the boundary of different land use types. Second, social media users are not evenly distributed between socioeconomic and demographic groups. For instance, Microblog users comprise mostly younger generations; thus, children and the elderly may be excluded from the study. Third, compared to traditional methods such as questionnaires and interviews, it is hard to obtain personal and subjective views from social media data. Ries et al. (2009) indicated that efforts to promote park use should increase awareness of park availability, improve perceptions of park quality and utilize social networks [9]. Social media data are lacking on the "awareness" part. Fourth, social media data generate results on overall trend or patterns while ignoring individual differences. Qualitative methods (e.g., in-depth individual interviews, focus group interviews, direct observation and participant observation) rely heavily on interpretations from participant language and actions and involve purposeful sampling of participants and settings. The contextualization in qualitative research helps to show such individual attributes. Data such as the Baidu vocabulary cloud may help overcome this weakness. Qualitative research may also be less amenable to standardized research procedures and more difficult to synthesize.

5. Conclusions

Urban parks are essential to the wellbeing of humans and are important places for the public to interact with nature. The use of urban parks is a dynamic process. For a single space, various activities may take place in different times of day, month, season and year. Exploring the determinants of urban parks utilization is helpful for improving the renewal strategy of urban parks, providing elastic space design schemes, promoting citizens' recreation, improving public physical and mental health and encouraging comprehensive benefits of urban parks. The use of new technique, new data and new analysis methods provide guidance for future renewal planning of urban parks, transportation and land use. The proposed methodology can be applied in similar high-density urban areas all over the world to enhance social, ecological, cultural and economic service functions of urban parks.

This study explores the determinants of urban parks utilization using microblog check-in data. Logarithmic linear regression models are constructed based on social media data, park basic data and social life data. Taking 17 comprehensive parks in central Shanghai as the research objects, this paper examines determinants affecting park utilization and puts forward corresponding optimization countermeasures from design and planning perspectives based on the results from regression models.

Our findings suggest that spring and autumn are peak seasons, while summer and winter are slack seasons. Seasonal climate has an apparent impact on park utilization. The frequency of visitors' park recreational activities is high in spring and autumn and low in summer and winter. When planning and designing recreational space for visitors, more emphasis should be placed on the design of various green outdoor recreational spaces, and the planning and allocation of sufficient recreational facilities to satisfy the peak demands.

Regarding the determinants of urban parks utilization, we find that both internal and external factors of parks contribute to urban park utilization. Participation in physical exercise is in high demand by visitors as a result of the social advocacy for a healthy life. In addition to sports venues, more elastic spaces should be designed to encourage diverse healthy activities, such as square dancing, Tai Chi, walking and even sitting. In addition, the inner quality of a park should be improved by adding attractive views such as water bodies, thoughtfully arranged plant scenes, comfortable sitting areas and social events in order to make parks more popular and improve their ranking. For external factors, we suggest enhancing surrounding transportation facilities and promoting mixed land use (residential areas, parks, commercial areas, other public service lands, etc.) for better urban parks utilization. Breaking the boundary between the city and parks, as well as increasing subway stations and bus stations to parks, can make parks more accessible and perceived as nice choices.

Author Contributions: Conceptualization, D.C.; methodology, Z.L. and D.C.; formal analysis, X.L.; writing—original draft preparation, D.C. and X.L.; writing—review and editing, C.L., D.C. and C.X.; supervision, S.C. and Z.L.; funding acquisition, S.C. and D.C. All authors have read and agreed to the published version of the manuscript.

Funding: This research was funded by National Natural Science Foundation of China, grant number 32001361.

Institutional Review Board Statement: Not applicable.

Informed Consent Statement: Informed consent was obtained from all subjects involved in the study.

Data Availability Statement: The data presented in this study are available on request from the corresponding author. The data are not publicly available due to privacy.

Conflicts of Interest: The authors declare no conflict of interest.

References

1. Woolley, H. *Urban Open Spaces*; Spon Press: London, UK; New York, NY, USA, 2003; pp. 25–36.
2. Mccormack, G.R.; Rock, M.; Toohey, A.M.; Hignell, D. Characteristics of urban parks associated with park use and physical activity: A review of qualitative research. *Health Place* **2010**, *16*, 712–726. [CrossRef]
3. Akpinar, A. How is quality of urban green spaces associated with physical activity and health? *Urban For. Urban Green.* **2016**, *16*, 76–83. [CrossRef]
4. Moreira, T.C.L.; Polizel, J.L.; Santos, I.; Fiilho, D.F.; Bensenor, I.; Lotufo, P.A.; Mauad, T. Green Spaces, Land Cover, Street Trees and Hypertension in the Megacity of Saõ Paulo. *Int. J. Environ. Res. Public Health* **2020**, *17*, 725. [CrossRef]
5. Chen, Y.; Liu, X.; Gao, W.; Wang, R.Y.; Li, Y.; Tu, W. Emerging social media data on measuring urban park use and their relationship with surrounding areas—A case study of shenzhen. *Urban For. Urban Green.* **2018**, *31*, 130–141. [CrossRef]
6. Waits, J. Urban green space: Is it the next financial frontier? *Bus. Perspect.* **2008**, *19*, 36–39.
7. Garvin, A.; Brands, R.M. *Public Parks: The Key to Livable Communities*, 1st ed.; W.W. Norton & Co.: New York, NY, USA, 2011; pp. 112–124.
8. Besenyi, G.M.; Kaczynski, A.T.; Stanis, S.A.; Bergstrom, R.D.; Lightner, J.S.; Hipp, J.A. Planning for health: A community-based spatial analysis of park availability and chronic disease across the lifespan. *Health Place* **2014**, *27*, 102–105. [CrossRef]
9. Ries, A.V.; Voorhees, C.C.; Roche, K.M.; Gittelsohn, J.; Yan, A.F.; Astone, N.M. A quantitative examination of park characteristics related to park use and physical activity among urban youth. *J. Adolesc. Health* **2009**, *45* (Suppl. S3), S64–S70. [CrossRef] [PubMed]
10. Golicnik, B.; Ward Thompson, C. Emerging relationships between design and use of urban park spaces. *Landsc. Urban Plan.* **2010**, *94*, 38–53. [CrossRef]
11. Senetra, A.; Krzywnicka, I.; Mielke, M. An analysis of the spatial distribution, influence and quality of urban green space—A case study of the Polish city of Tcze. *Bull. Geogr. Socio-Econ. Ser.* **2018**, *42*, 129–149. [CrossRef]
12. Giles-Corti, B.; Broomhall, M.H.; Knuiman, M.; Collins, C.; Douglas, K.; Ng, K.; Lange, A.; Donovan, R.J. Increasing walking: How important is distance to, attractiveness, and size of public open space? *Am. J. Prev. Med.* **2005**, *28* (Suppl. S2), 176.
13. Whyte, W.H. *The Social Life of Small Urban Spaces*; Conservation Foundation: Washington, DC, USA, 1980; pp. 42–56.
14. Mccormack, G.R.; Rock, M.; Swanson, K.; Burton, L.; Massolo, A. Physical activity patterns in urban neighborhood parks: Insights from a multiple case study. *BMC Public Health* **2014**, *14*, 962. [CrossRef] [PubMed]
15. Kwan, C.P.; Jim, C.Y. Subjective outdoor thermal comfort and urban green space usage in humid-subtropical hong kong. *Energy Build.* **2018**, *173*, 150–162.

16. Hawelka, B.; Sitko, I.; Beinat, E.; Sobolevsky, S.; Kazakopoulos, P.; Ratti, C. Geo-located twitter as proxy for global mobility patterns. *Cartogr. Geogr. Inf. Sci.* **2014**, *41*, 260–271. [CrossRef] [PubMed]
17. Steiger, E.; de Albuquerque, J.P.; Zipf, A. An advanced systematic literature review on spatiotemporal analyses of twitter data. *Trans. GIS* **2015**, *19*, 809–834. [CrossRef]
18. Jendryke, M.; Balz, T.; Mcclure, S.C.; Liao, M. Putting people in the picture: Combining big location-based social media data and remote sensing imagery for enhanced contextual urban information in shanghai. *Comput. Environ. Urban Syst.* **2017**, *62*, 99–112. [CrossRef]
19. Noulas, A.; Scellato, S.; Lambiotte, R.; Pontil, M.; Mascolo, C. A tale of many cities: Universal patterns in human urban mobility. *PLoS ONE* **2011**, *7*, e37027.
20. Zhai, S.; Xu, X.; Yang, L. Mapping the popularity of urban restaurants using social media data. *Appl. Geogr.* **2015**, *63*, 113–120. [CrossRef]
21. Livia, H.; Ross, P. Exploring place through user-generated content: Using Flickr to describe city cores. *J. Spat. Inf. Sci.* **2010**, *1*, 21–48.
22. García-Palomares, J.; Gutiérrez, J.; Mínguez, C. Identification of tourist hot spots based on social networks: A comparative analysis of European metropolises using photo-sharing services and gis. *Appl. Geogr.* **2015**, *63*, 408–417. [CrossRef]
23. Fujisaka, T.; Lee, R.; Sumiya, K. Exploring urban characteristics using movement history of mass mobile microbloggers. *Elev. Workshop Mob. Comput. Syst. Appl.* **2010**, *10*, 13–18.
24. Kennedy, L.; Naaman, M.; Ahern, S.; Nair, R.; Rattenbury, T. How flickr helps us make sense of the world: Context and content in community-contributed media collections. In Proceedings of the 15th ACM International Conference on Multimedia, Augsburg, Bavaria, Germany, 23–28 September 2007; pp. 631–640.
25. Friasmartinez, V.; Soto, V.; Hohwald, H.; Friasmartinez, E. Characterizing Urban Landscapes Using Geolocated Tweets. In Proceedings of the 2012 International Conference on Privacy, Security, Risk and Trust and 2012 International Confernece on Social Computing, Amsterdam, The Netherlands, 3–5 September 2012.
26. Roberts, H.V. Using twitter data in urban green space research: A case study and critical evaluation. *Appl. Geogr.* **2017**, *81*, 13–20. [CrossRef]
27. Agryzkov, T.; Martí, P.; Nolasco-Cirugeda, A.; Serrano-Estrada, L.; Tortosa, L.; Vicent, J.F. Analysing successful public spaces in an urban street network using data from the social networks foursquare and twitter. *Appl. Netw. Sci.* **2016**, *1*, 12. [CrossRef]
28. Roberts, H.; Sadler, J.; Chapman, L. Using twitter to investigate seasonal variation in physical activity in urban green space. *Geo Geogr. Environ.* **2017**, *4*, e00041. [CrossRef]
29. Xu, W. Urban explorations: Analysis of public park usage using mobile GPS data. In Proceedings of the Bloomber Data for Good Exchange Conference, Chicago, IL, USA, 24 September 2017.
30. Manaugh, K.; Kreider, T. What is mixed use? Presenting an interaction method for measuring land use mix. *J. Transp. Land Use* **2013**, *6*, 63–72. [CrossRef]
31. Crompton, J.L. The impact of parks on property values: Empirical evidence from the past two decades in the United Sates. *Manag. Leis.* **2005**, *10*, 203–218. [CrossRef]
32. Dunse, N.; White, M.; Dehring, C. Urban Parks, Open Space and Residential Property Values. *RICS Res. Pap. Ser.* **2007**, *7*, 9–36.
33. Chen, C.; Luo, W.; Li, H.; Zhang, D.; Xia, Y. Impact of perception of green space for health promotion on willingness to use parks and actual use among young urban residents. *Int. J. Environ. Res. Public Health* **2020**, *17*, 5560. [CrossRef]
34. Shen, B.L. Study on the eutrophication evaluation for small landscape waters in cities. *J. Anhui Agric. Sci.* **2011**, *39*, 14321–14325.
35. Stoianov, I.; Chapra, S.; Maksimovic, C. A framework linking urban park land use with pond water quality. *Urban Water* **2000**, *2*, 47–62. [CrossRef]
36. Wang, Y.C.; Lin, J.C.; Liu, W.Y.; Lin, C.C.; Ko, S.H. Investigation of visitors' motivation, satisfaction and cognition on urban forest parks in taiwan. *J. For. Res.* **2016**, *21*, 261–270. [CrossRef]
37. Frank, S.; Fürst, C.; Koschke, L.; Witt, A.; Makeschin, F. Assessment of landscape aesthetics—Validation of a landscape metrics-based assessment by visual estimation of the scenic beauty. *Ecol. Indic.* **2013**, *32*, 222–231. [CrossRef]
38. Gui, Y. Recreation Demand Preference Research of Citizens in Shanghai. *Shanghai Urban Plan. Rev.* **2016**, *3*, 102–108.
39. Liao, C.; Brown, D.; Fei, D.; Long, X.; Chen, D.; Che, S. Big data-enabled social sensing in spatial analysis: Potentials and pitfalls. *Trans. GIS* **2018**, *22*, 1351–1371. [CrossRef]

Article

Concatenated Residual Attention UNet for Semantic Segmentation of Urban Green Space

Guoqiang Men [1,2], Guojin He [1,2,3,*] and Guizhou Wang [1,3]

[1] Aerospace Information Research Institute, Chinese Academy of Sciences, Beijing 100094, China; menguoqiang19@mails.ucas.ac.cn (G.M.); wanggz@aircas.ac.cn (G.W.)
[2] College of Resource and Environment, University of Chinese Academy of Sciences, Beijing 100049, China
[3] Satellite Remote Sensing Technology Department, Key Laboratory of Earth Observation Hainan Province, Sanya 572029, China
* Correspondence: hegj@aircas.ac.cn; Tel.: +86-010-8217-8188

Citation: Men, G.; He, G.; Wang, G. Concatenated Residual Attention UNet for Semantic Segmentation of Urban Green Space. *Forests* **2021**, *12*, 1441. https://doi.org/10.3390/f12111441

Academic Editors: Manuel Esperon-Rodriguez and Tina Harrison

Received: 3 October 2021
Accepted: 21 October 2021
Published: 22 October 2021

Abstract: Urban green space is generally considered a significant component of the urban ecological environment system, which serves to improve the quality of the urban environment and provides various guarantees for the sustainable development of the city. Remote sensing provides an effective method for real-time mapping and monitoring of urban green space changes in a large area. However, with the continuous improvement of the spatial resolution of remote sensing images, traditional classification methods cannot accurately obtain the spectral and spatial information of urban green spaces. Due to complex urban background and numerous shadows, there are mixed classifications for the extraction of cultivated land, grassland and other ground features, implying that limitations exist in traditional methods. At present, deep learning methods have shown great potential to tackle this challenge. In this research, we proposed a novel model called Concatenated Residual Attention UNet (CRAUNet), which combines the residual structure and channel attention mechanism, and applied it to the data source composed of GaoFen-1 remote sensing images in the Shenzhen City. Firstly, the improved residual structure is used to make it retain more feature information of the original image during the feature extraction process, then the Convolutional Block Channel Attention (CBCA) module is applied to enhance the extraction of deep convolution features by strengthening the effective green space features and suppressing invalid features through the interdependence of modeling channels.-Finally, the high-resolution feature map is restored through upsampling operation by the decoder. The experimental results show that compared with other methods, CRAUNet achieves the best performance. Especially, our method is less susceptible to the noise and preserves more complete segmented edge details. The pixel accuracy (PA) and mean intersection over union (MIoU) of our approach have reached 97.34% and 94.77%, which shows great applicability in regional large-scale mapping.

Keywords: urban green space; remote sensing; deep learning; convolutional neural network; residual structure; attention mechanism

1. Introduction

Regarding the concept of urban green space, different regions have their own interpretation of its definition and scope. Compared with urban green space, western countries use the concept of urban open space more in land use planning [1–3]. Urban open space is an open space area reserved for parks and other "green spaces", which includes water and other natural environments in addition to vegetation [4]. In China, in order to standardize the management process of urban greening, the government has issued the "Urban Green Space Classification Standard", which divides urban green space into two parts, including green space within urban construction land and square land and regional green space outside urban construction land [5]. In this study, the above definition of urban green space is followed.

Urban green space is an indispensable element in the urban ecosystem which is always considered to be an important component to improve the quality of the urban ecological environment [6]. It provides protection for the sustainable development of the city in various aspects of ecological service functions, such as reducing greenhouse gases, regulation of urban climate, reduction of energy consumption, maintenance of ecological security, etc. [7–10]. However, with the rapid development of urbanization, urban built-up areas continue to expand, and green spaces are severely damaged, affecting the quality of life of residents. The unreasonable planning and construction of urban green space restrict the healthy development of the city. Therefore, good urban green space monitoring is a necessity for the sustainable development and management of cities [11]. How to accurately and dynamically obtain urban green space information has arisen the interest of researchers.

Since the 21st century, with the rapid development of Earth observation technology, the acquisition ability of satellite remote sensing data has been greatly improved, suggesting that it has entered a new era of multi-platform, multi angle, multi-sensor, all-time and all-weather Earth observation. Remote sensing technology can quickly and accurately monitor the dynamic changes of green space in a study area, so itis suitable for large-scale resource investigation and research. At present, a variety of optical and radar remote sensing data sources including Landsat series data [12], GaoFen series data [13] and Sentinel series data [14] have been employed for urban green space extraction. For the application of remote sensing technology in urban green space research, the key technology is how to quickly and accurately obtain the surface vegetation coverage information. Using remote sensing image to extract urban green space is to identify and classify the types of land cover, so as to obtain the vegetation cover map of the real situation of land cover. The methods for urban green space extraction from remote sensing images can be divided into four kinds: threshold method [15,16], pixel-based classification method represented by machine learning [17–20], object-oriented classification method [21,22] and deep learning method [23–26]. The threshold method selecta an appropriate threshold to distinguish the green space through the difference of the spectral response of vegetation and other ground objects in one or more bands. Due to the difference between the reflection of vegetation in the visible and near-infrared bands and the soil background, the researchers improved the combination of bands and proposed a series of vegetation indices to extract surface vegetation coverage information [16]. However, because of the different complexity of the environmental background in urban areas, the representation of vegetation on remote sensing images is easily disturbed by other features in the built-up area, especially in the classification of vegetation in areas with high-density buildings in cities. With the popularity of machine learning technology, more machine learning-based algorithms have attracted great attention from researchers, support vector machines [17], decision trees [18], random forests [19], artificial neural network [20] and other algorithms. These approaches are widely used in the research of urban green space extraction. However, there are a large number of similar objects with different spectra in high-resolution remote sensing images, which makes the traditional pixel-based classification methods unable to accurately distinguish different types of ground objects. As the traditional methods have become mature, it is difficult to improve at the technical level. In order to improve accuracy, traditional urban green space machine learning algorithms require manual design of features including texture and terrain [27]. This process is time-consuming and laborious, and there are certain challenges for accurately extracting urban green space information.

Deep learning has become a popular method for remote sensing information extraction. The principle of applying deep learning technology to urban green space information extraction is that the most original remote sensing image passes through multiple hidden layers and abstracts from low level to high level, aiming to automatically select the characteristics of the target to discover the distributed feature representation of green space. [28]. Compared with traditional machine learning, the advantage of deep learning is that without manual design and acquisition of features, unsupervised or semi-supervised feature learning and efficient hierarchical feature extraction algorithms are implemented for feature extraction [29]. The convolutional neural network (CNN) is one of the most successful network architectures in deep learning algorithm which consists of input layer, convolution layer, pooling layer, full connection layer and output layer [30]. The input layer is used to input the original data; the convolution layer is used for feature extraction; the pooling layer compresses the input feature map; the full connection layer connects all features and sends the output value to the classifier; and the output layer outputs the classification result. The reasonable architecture ensures that the feature learning process is efficient, so it has become the main algorithm to extract urban green space coverage information. Nijhawan proposed a framework that combines local binary pattern (LBP) and GIST features with multiple parallel CNNs for feature extraction, and then combined with SVM to extract vegetation in the city. As the number of parallel CNNs increases, the accuracy increases significantly [23]. Moreno-Armendáriz built a deep neural network system based on CNN and multi-layer perceptron (MLP) to evaluate the health of urban green spaces and promote the realization of the sustainable development goals of smart cities [24]. Timilsina proposed an object-based convolutional neural network (OB-CNN) for extracting the coverage changes of the number of cities with an accuracy of over 95%, indicating that object-based CNN technology can effectively achieve urban tree coverage Mapping and monitoring [25]. Hartling tested the ability of densely connect convolutional networks (DenseNet) to identify the main tree species in complex urban environments in the fusion image of WorldView-2 Visible-to-Near Infrared (VNIR), Worldview-3 Selective Wavelength Infrared (SWIR) and LiDAR data sets. The study showed that, regardless of the size of the training sample, DenseNet is superior to RF and SVM technologies when processing highly complex image scenes, so it is more effective for urban tree species classification [26].

Compared with the above methods, urban green space is not only often blocked by shadows on high-resolution remote sensing images, but misclassified owing to the spectrum similarity of farmland, or-chards, etc. Therefore, it is difficult to extract urban green space using only spectral information. At the same time, due to the complex background of the ground features and irregular boundaries, the existing methods will produce some misclassifications and omissions during extraction. Therefore, in order to solve the above problems, this paper proposes an improved fully convolutional neural network based on the encoding and decoding structure to extract urban green space from the Gaofen-1 remote sensing image which called concatenated residual attention UNet (CRAUnet). The work presented in this article focuses on the following three aspects: (1) A residual module with feature concatenation mechanism is proposed to improve the loss of original image features. (2) In order to improve the feature expression ability of the network, attention mechanism is embedded to the model, and convolutional block channel attention (CBCA) module is proposed. (3) To illustrate the applicability of the network, we compare with other classical networks to evaluate the efficiency of the network structure.

2. Materials and Methods

2.1. Study Area and Data Sources

As a national economic center city and an international city, Shenzhen is China's first Special Economic Zone [31]. Its geographical location is shown in Figure 1. As one of the first cities to develop and endure environmental pressure earlier, Shenzhen has always adhered to both protection and development, and vigorously created a good forest ecological environment. Especially since 2015, Shenzhen has fully initiated the creation of a national forest city. By 2019, the green area coverage exceeded 45% [32].

The Gaofen-1 satellite was successfully launched into orbit on 26 April 2013. The selected panchromatic multispectral camera image includes a panchromatic band with a spatial resolution of 2 m and four multispectral bands with a spatial resolution of 8m. In this research, we selected 12 scenes multi temporal Gaofen-1, 1-C and 1-D remote sensing data of Shenzhen for 2017 and 2020, of which 9 scenes are used as training data and 3 scenes are used as testing data. The information about images of the study is shown in Table 1. The PCI Geo Imaging Accelerator software was employed to conduct the remote sensing data preprocessing. The rational polynomial coefficient model was adopted to geometrically correct the remote sensing image [33], then the PANSHARP method was employed to fuse the multi-spectral image and the pan-chromatic image [34], and finally all the data was unified into 8-bit. After the preprocessing, the geometric error of the remote sensing image is limited to within 1 pixel. The flow chart of preprocessing is shown in Figure 2.

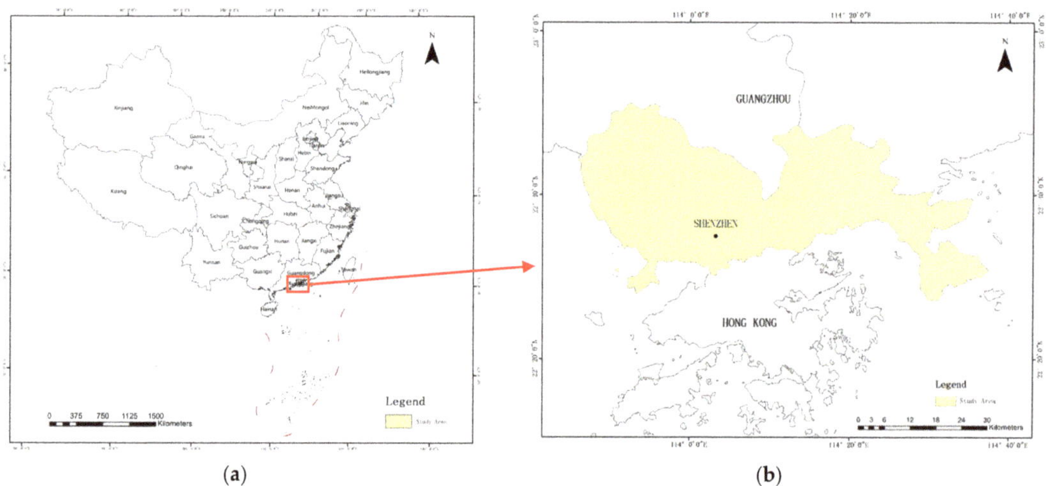

(**a**)   (**b**)

Figure 1. Schematic diagram of the geographical location of the study area. (**a**) Shenzhen's location in China. (**b**) Overview of Shenzhen.

Table 1. Detailed information of the study data.

Satellite	Central Longitude (°E)	Central Latitude (°N)	Imaging Time	Image Size (Pixel × Pixel)	Image Usage
Gaofen-1	114.2	22.4	11 December 2017	21,196 × 21,103	Training
Gaofen-1	114.2	22.7	11 December 2017	21,200 × 21,106	Training
Gaofen-1	113.7	22.7	19 December 2017	20,599 × 20,456	Training
Gaofen-1	113.9	22.4	15 February 2017	23,436 × 22,351	Training
Gaofen-1	114.0	22.6	15 February 2017	23,976 × 22,962	Training
Gaofen-1	114.5	22.4	11 December 2017	21,476 × 21,347	Testing
Gaofen-1	114.6	22.7	11 December 2017	21,470 × 21,348	Testing
Gaofen-1B	114.1	22.4	22 February 2020	41,008 × 40,888	Training
Gaofen-1B	114.2	23.0	22 February 2020	41,008 × 40,890	Training
Gaofen-1B	114.6	22.4	12 January 2020	40,936 × 40,837	Training
Gaofen-1C	113.7	22.4	26 October 2020	40,003 × 39,699	Training
Gaofen-1C	113.9	23.0	26 October 2020	39,986 × 39,678	Testing

2.2. Methods

2.2.1. Sample Generation

Since deep learning requires a large amount of labelled data related to the classification target for training, and the existing open-source datasets cannot meet the requirements of this article, it was necessary to manually establish an urban green space labelled dataset. Generally, the boundary of green space in the remote sensing image of the experimental area is directly delineated by manual visual interpretation. However, due to the fragmented distribution, diverse types and complex background of green space, this method is time-consuming and laborious [22]. Therefore, this paper adopts the object-oriented method for data annotation. The process mainly realized through three steps including selecting appropriate parameters to segment the image, classifying the segmented image, and manually correcting the result. The final result is stored as 8-bit ground truth binary label, where the value 1 represents the urban green space while 0 denotes the background. The sample data covers different types of green space, as shown in Figure 3. Then the processed remote sensing image data and label data are divided into training set and test set according to the area ratio of 4:1 to ensure that the training set and the test set are independent and identically distributed. At the same time, to increase the diversity of training samples and reduce the under-fitting problem caused by insufficient samples during the training process, random cropping strategy is utilized to generate the samples with the size of 256 × 256 pixels by selecting different repetition rates, and random enhancement is adopted ed to expand the training set and the test set to weaken the background noise and enhance the robustness of the model. In this paper, we mainly use flip, rotation, scaling, adding Gaussian noise and fuzzy to enhance the samples, and finally there are in total 8000 training samples and 2000 test samples, respectively.

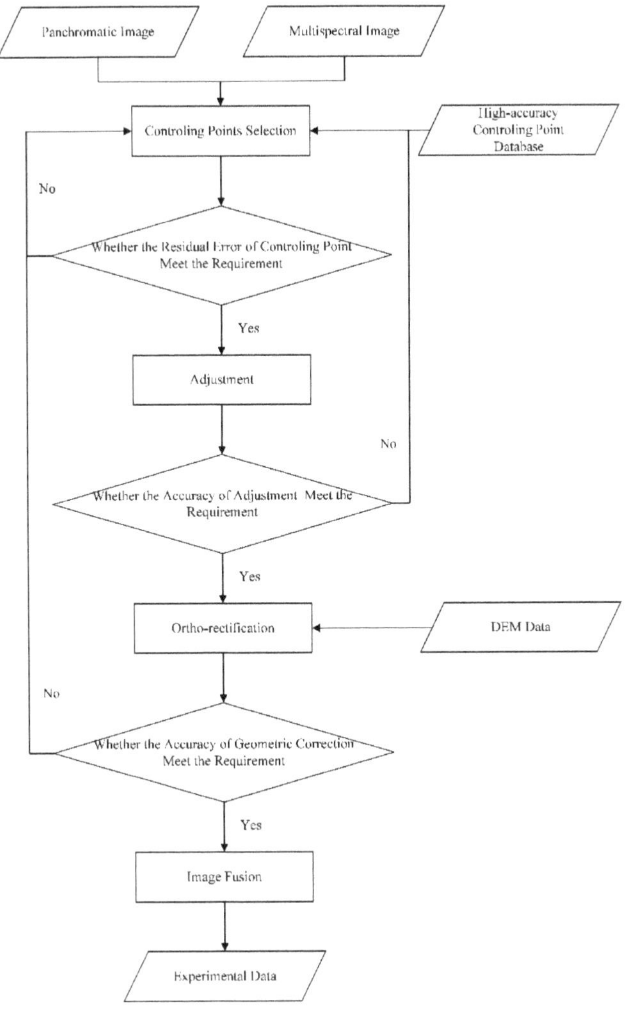

Figure 2. Flow chart of remote sensing image preprocessing.

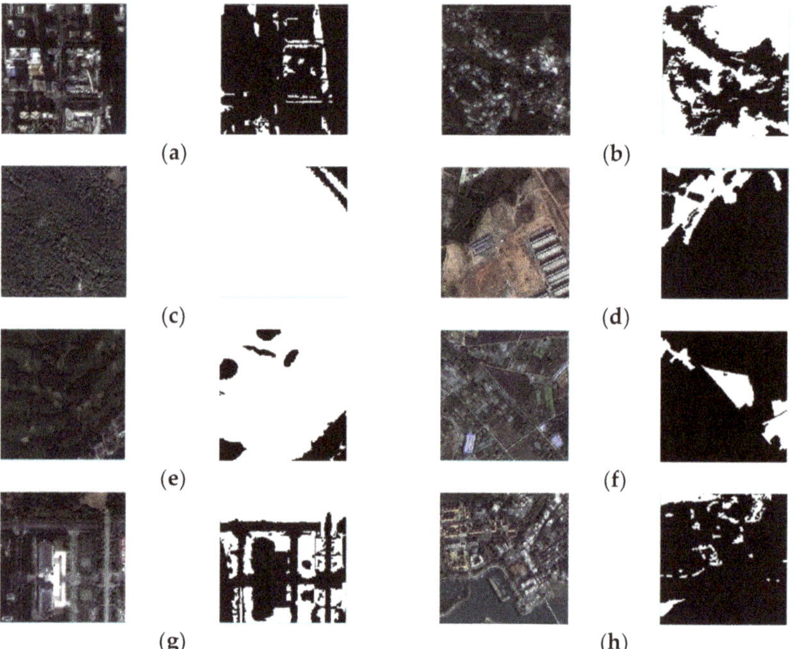

Figure 3. For samples of different green space types, the left side of each group is the original image and the right side is the label image. (**a**) High density building area; (**b**) countryside; (**c**) Suburban Forest; (**d**) Bare land; (**e**) Golf Course; (**f**) Farmland; (**g**) Land for parks and squares; (**h**) Port.

2.2.2. Improved Residual Structure

In deep learning, based on the network training process of stochastic gradient descent, the error signal is prone to gradient dispersion or gradient explosion through multi-layer back propagation [35]. Therefore, as the network depth increases, training becomes more difficult. In order to solve the above degradation phenomenon, He and others proposed residual neural network (ResNet) in 2015, which greatly eliminated the problem of training difficulties caused by the excessive depth of the neural network [36]. The main contribution of ResNet is to invent a building block containing direct mapping for the phenomenon of network degradation [37]. This process can be expressed as follows:

$$x_{l+1} = x_l + F(x_l, w_l)\, x_l \tag{1}$$

where x_l and x_{l+1} are input and output of the *l-th* unit, and F is a residual function.

The residual building block is divided into an identity mapping part and a residual part, as shown in Figure 4a. x_l is the identity map which is reflected in the straight line on the left side of the Figure 4a. In the convolution neural network, the number of feature maps of x_l and x_{l+1} may be different. At this time, one needs to use a 1×1 convolution for dimensionality increase or dimensionality reduction. $F(x_l, w_l)$, presented at the right side of the Figure 4a, represents the residual part, which is usually composed of two or three convolution operations. The residual structure can alleviate the gradient dispersion problem of deep neural network to a certain extent, solve the degradation problem of deep neural networks, which can make the forward and backward propagation of information smoother.

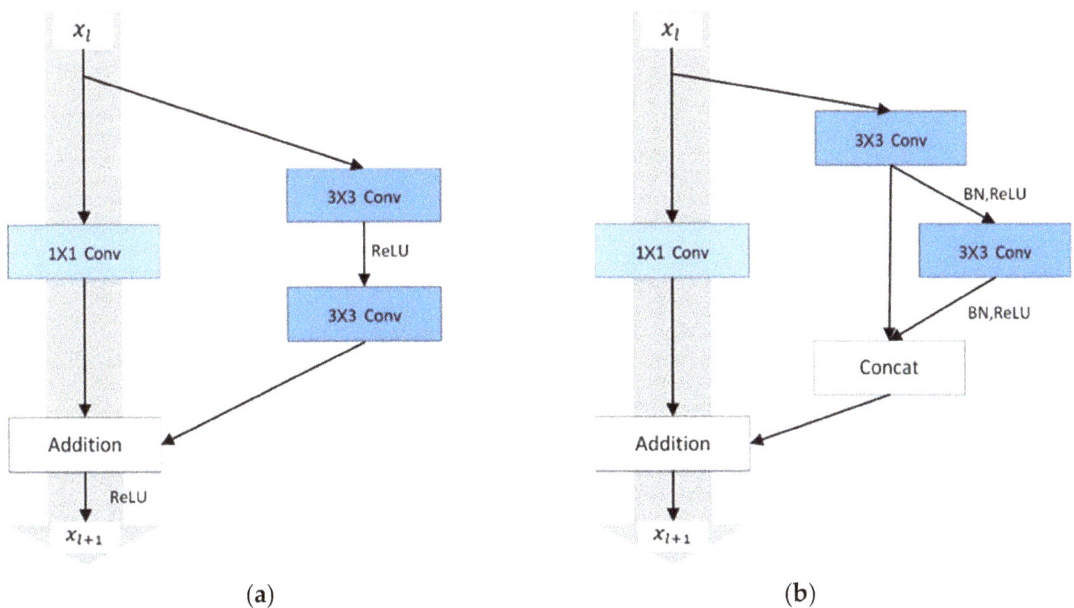

Figure 4. Comparison of different residual structures. (**a**) The original residual structure; (**b**) The proposed residual structure.

Different from other ground objects, the types of urban green space are diverse and its distribution is highly fragmented. At the same time, they are often misclassified as a result of shadow occlusion in remote sensing images and the similarity of spectral features with cultivated land usually leads to confusion between targets, so effective use of any feature of the original image is particularly important for the accurate extraction of information. Therefore, it is necessary to introduce the residual structure into the extraction of urban green space to solve the above problems. In this research, we propose a new residual structure based on the idea of identity mapping in residual neural network, as shown in Figure 4b. This building block passes through a convolutional layer with a kernel size of 1×1 in the identity mapping part to increase the number of features to the required output feature size. The 1×1 convolution kernel is used because it summarizes the features between pixels with a smaller kernel to avoid the loss of initial image information. In the residual part, in order to make full use of different levels of semantic features, while reducing the amount of network parameters and improving computational efficiency, the input image is convolved by halving the number of two consecutive features. Then the output results of each convolution are concatenated to obtain the semantic information of urban green space from multiple feature dimensions. At the same time, batch norm (BN) is used as normalization after each convolution kernel, and then rectified linear unit (ReLU) is used as activation function.

2.2.3. Convolutional Block Channel Attention Mechanism

In the convolutional neural network, each convolutional layer contains several filters, which can learn to express the local spatial connection pattern including all channels. However, the local channel features of adjacent spatial locations in the feature map often have a high correlation due to their overlapping receptive fields. The attention mechanism in deep learning originates from the attention mechanism of human brain. When the human brain receives external information, such as visual information and auditory information, it often does not process and understand all information, but only focuses on some significant or interesting information, which is conducive to filtering out unimportant information to improve the efficiency of information processing [38]. Introducing the attention mechanism into the convolutional neural network does not significantly increase the amount of network parameters while making selective and finer adjustments on the existing feature maps to improve the performance of the network and increase the interpretability of the neural network structure [39]. Therefore, in this study, to emphasize the spectral relationship of remote sensing image feature map, we introduce channel attention mechanism and propose a new network unit—a convolution block channel attention (CBCA) module—to replace the skip connection in the original UNET structure. The structure of the module is shown in Figure 5.

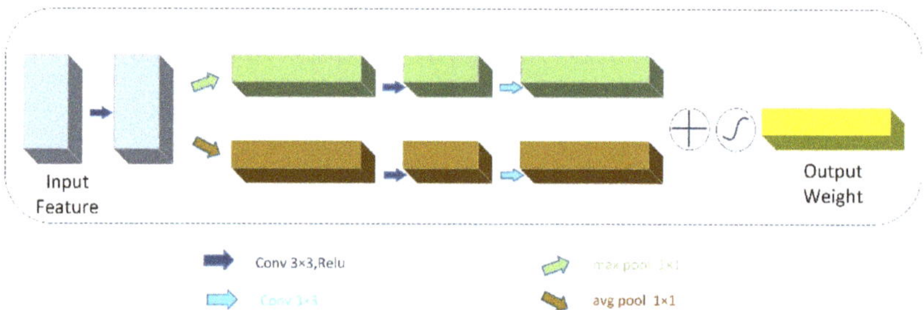

Figure 5. The structure of the convolution block channel attention module.

First, the feature map output by the residual module is subjected to a standard convolution to further enhance the feature expression ability. Then, the pooling layer is used to compress each channel into a single-dimensional digital vector. Here, we use max pooling and average pooling to compress the feature map in the channel dimension to obtain two different channel feature descriptors. Then, the above results are input into the fully connected layer with ReLU activation to reduce the dimension and introduce new nonlinearity, and finally a sigmoid activation function is employed to provide each channel a smooth weight, and output the weight value. The range of values is normalized to be between 0–1. Finally, the weights are multiplied by the features of the original encoder output, and then transferred to the decoder. The process can be described by the following formula:

$$\lambda_{weight} = \sigma\left(F_{FC}\left(P_{avg}\left(F_{3\times3}(x_l)\right)\right) + F_{FC}\left(P_{max}\left(F_{3\times3}(x_l)\right)\right)\right) \tag{2}$$

where σ, F_{FC}, and P represent the *Sigmoid*, *FC*, and *Pooling Operation*, respectively.

2.2.4. Network Structure

In this paper, a convolution neural network for extracting urban green space is proposed, we named it concatenated residual attention UNet (CRAUNet). The network chooses UNet as the backbone of the network, that is, using the encoder and decoder structure combined with shallow semantic information to achieve a smooth transformation from the image to the segmentation mask [40]. As shown in Figure 6, the architecture of the network can be divided into the following three parts: encoder, skip connection with attention mechanism, and decoder.

In the encoder part, in order to eliminate the gradient dispersion and explosion phenomenon in the deep neural network and make the forward and backward propagation of information smoother, this part of the building block is replaced with an improved convolution residual module. In addition to the residual structure proposed in this paper, this module also includes a max pooling layer after it. The size of the convolution kernel of the convolution operation is 3×3, followed by a BN layer and a ReLU. The max pooling layer downsamples the input image to half of the original size. After each encoder building block, the number of input feature channels will be doubled, and the image size will be halved. The structure of the decoder is similar to that of the encoder. The residual structure is not adopted and the maximum pooling layer is replaced with an upsampling layer. Here, we use the bilinear interpolation method to finally restore the feature map to its original size. In each decoder building block, to ensure that the output is the same size as the encoder output, the upsampling rate is set to 2. In the skip connection part, in order to make better use of the information of the multi-resolution feature map, the CBCA mode is embedded to enhance the useful green space features in the channel dimension and suppress the invalid background features, thereby improving the computational efficiency of the network model. Then, the output of this part is concatenated with the output of the previous layer of the decoder and input to the decoder of the current layer. The process is formulated as follows:

$$D_i = F_{Conv}\left(F_{up}D_{i+1}\right) + \lambda_{weight}E_i \tag{3}$$

where D_i indicates the hidden feature of decoder part in i layers, D_{i+1} denotes the lower hidden feature of D_i, and F_{up} represents the up-sampling operation.

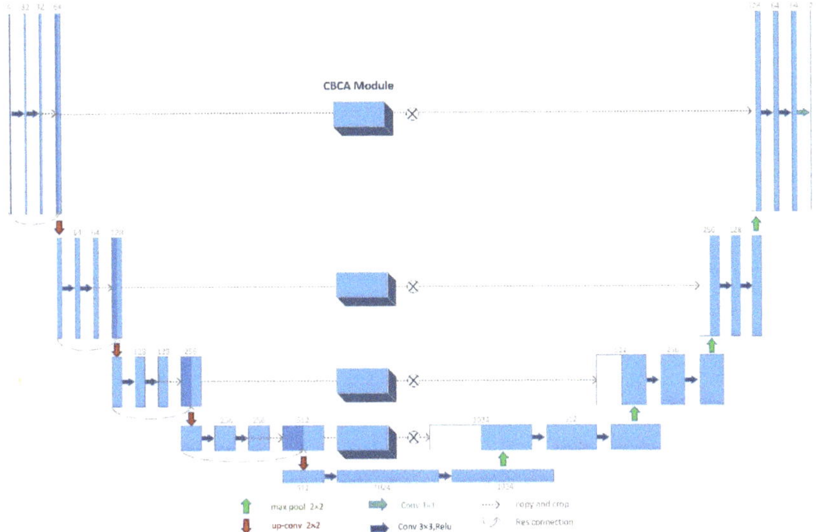

Figure 6. The structure of the concatenated residual attention UNet.

This approach for increasing the resolution of the convolution features can provide more fine features for segmentation on the region of the edge and avoid chessboard effect [41].

2.2.5. Inference Methodology

The advantage of a fully convolutional neural network is that it can make full use of the contextual information of the image to improve its performance. However, due to the large size of the high-resolution remote sensing image, directly inputting the data without preprocessing is likely to cause insufficient graphics card memory. Therefore, before the large-scale remote sensing images are input into the network architecture, the images need to be cropped into several smaller images of fixed size and predicted, respectively. After the results of each smaller tile are obtained, the whole images predicted by the network architecture are obtained by stitching operation. However, due to the large number of convolution processing in the network model, the center pixel can obtain more context information while the edge pixels gain limited information, so the pixels close to the edge of the image will not be classified as accurately as the pixels close to the center. After stitching the results of several small images into the original image size, obvious stitching traces will appear in some positions [42]. In order to alleviate these problems, this experiment adopts the sliding window method in the network inference phase. In our experiments, a large-size remote sensing image is cropped into multiple small images of 256×256 pixels. Notably, during the cropping process, a fixed numerical area smaller than the image size is set to control the prediction area range of each input to the network frame. In this experiment, the size of the sliding window is set to 64.

2.2.6. Accuracy Assessment

In this study, seven evaluation indexes were selected to evaluate the performance, including pixel accuracy (PA), the precision, the recall, F_1-Score (F_1), and the mean intersection over union (MIoU), floating point of operations (FLOPs). The definitions and formulas of these indicators are listed in Table 2.

Table 2. Evaluation metrics for the accuracy assessment.

Accuracy Evaluation Criteria	Definition	Formula
PA	The ratio of the correctly classified number of pixels and the total number of pixels	$PA = \frac{TP+TN}{TP+TN+FP+FN}$
Precision	The ratio of the number of correctly classified pixels and the number of the labeled pixels	$Precision = \frac{TP}{TP+FP}$
Recall	The ratio of the number of correctly classified pixels and the number of the actual target feature pixels	$Recall = \frac{TP}{TP+FN}$
IoU	The ratio of the intersection and the union of the ground truth and the predicted area	$IoU = \frac{TP}{TP+FN+FP}$
MIoU	The average IoU for all classes	$MIoU = \frac{1}{n+1} \sum_{i=0}^{n} IoU$
F_1-Score	The harmonic means for precision and recall	$F_1 = 2 \times \frac{Precision \times Recall}{Precision + Recall}$
Params	The total weight parameters of all parameterized layers of the model	
FLOPs	The number of multiplication and addition operations in the model	

Where TP FP FN and TN are the is true positive, false positive, false negative and true negative classifications, respectively.

3. Experiment

In this section, in order to evaluate the performance of the various modules used in the designed network on the constructed urban green space dataset, we conducted two ablation experiments. In terms of evaluation indicators, we mainly use pixel accuracy (PA), and present the number of parameters and FLOPs for each network. Because it is time-consuming to test the final performance of the network, we need to reduce the learning rate and fine tune repeatedly in the process, so we only care about the performance of different network structures in the fixed hyperparameters and epoch training.

3.1. Effectiveness of Concatenated Residual Structure

In the first experiment, we tested the convergence of the proposed residual structure. In order to ensure the consistency, we choose UNet as our baseline model without a dropout layer, and the bilinear interpolation method is selected for upsampling. Then, we modify the baseline model by adding identity mapping for each layer of the UNet network decoder part, thereby increasing the residual structure (ResUNet). It should be clear that we have added a 1×1 convolution kernel to the identity mapping part to ensure that the number of features in this part is increased to the required output feature size, so there areslightly more parameters than the Unet model. Then we reduce the number of network parameters by reducing the number of convolution channels in the residual block by half and connecting the output of each convolution kernel (CRUNet). At the same time, for comparative analysis, we also do the above operations on the baseline model without adding a residual structure (CUNet). In Figure 7, we plot the pixel accuracy curves of all models in the validation set. It can be seen that with the modification of each module, the performance difference of the convergence rate among the model is increasing. The baseline UNet needs about 120 epochs to reach the same level of performance that CRUNET achieved in 70 epochs. Only by modifying the convolution in the UNet decoder and concatenating the features of different convolution kernels, the convergence speed of the model can be significantly improved. In addition, it can be seen from the comparison that if only the residual module is added to the baseline model, the convergence speed of the model can be improved while the accuracy of the model will fluctuate more during the training process, and the training will be unstable. Moreover, we also compared the number of parameters and calculations of different network structures. As presented in Table 3, the introduced residual structure can improve the pixel accuracy while reducing the complexity of the model. The results show that when the convolution kernel of the decoder in the baseline network is modified and the residual module is added, the model can learn more features of urban green space, and the convergence performance of the model is significantly improved.

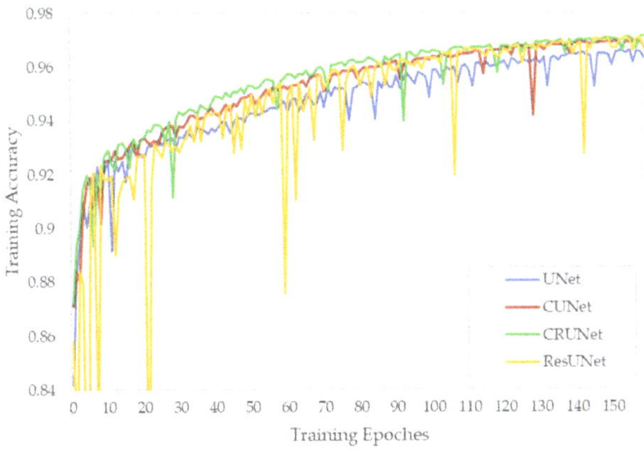

Figure 7. Convergence performance of the CRUNet architecture. Starting from a baseline UNet, we add components keeping all training hyperparameters identical.

Table 3. The efficiency comparison of the various methods.

Model	PA (%)	Params (M)	FLOPs (G)
UNet	96.5	31.04	46.20
CUNet	96.9	27.91	37.03
CRUNet	97.1	28.09	37.46
ResUNet	97.0	32.80	52.51

3.2. Performance of Convolution Block Channel Attention Module

In the second experiment, we explored the effect of introducing the attention mechanism into skip connections on the performance of network feature extraction. Similarly, we also adopted UNet as the baseline model of this experiment to illustrate the role played by the attention mechanism. Then, we use the channel attention module (CAM) and the CBCA module proposed in previous section to replace the original skip connection, and we also applied the CBCA module to the CRUNet model mentioned above. As can be seen from Figure 8, the attention mechanism is introduced into the model, which makes use of the relationship between channels to improve the feature extraction ability of the model and compress the redundant features on the channel dimension. After adding the attention mechanism, the performance of the model is improved from 96.5% to 97.1%, obtaining a rise of 0.6%. This demonstrates that, compared with the original skip connection, the skip connection with attention mechanism can extract green space features more effectively.

After feature fusion, it can recover high-resolution details better when upsampling, and the combination with residual structure can improve the convergence speed of the model.

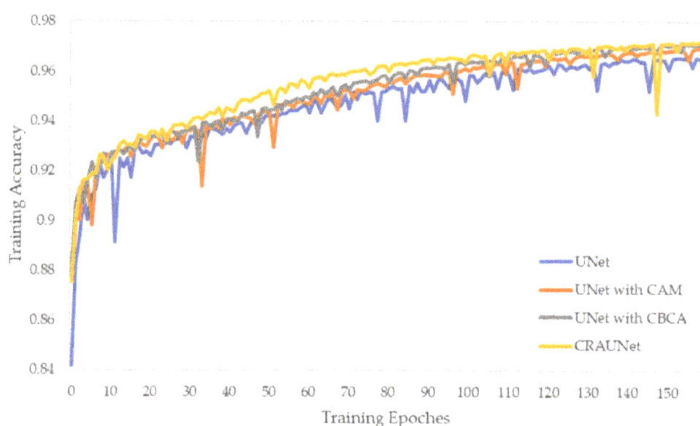

Figure 8. Convergence performance of the CBCA module. We compare two modules, a CAM and a CBCA module, and introduce the CBCA module to the CRUNet.

4. Results

In this section, we will explore the performance of the proposed model and compare the performance with other structures. The training process is implemented on an NVIDIA Titan V GPU using Python 3.6, PyTorch deep learning framework, accelerated by cuDNN 10.0. Aftereach training epoch, the classification accuracy on the training data set and the validation data set is calculated. Our goal is to gain the best convergence model to the best and apply it for the extraction of urban green space.

4.1. Comparative Experiment with Other Networks

In order to illustrate the performance of CRUNet, we selected several typical semantic segmentation networks including FCN8s, UNet, Deeplab for comparison in the urban green space data set. The experimental results are shown in Table 4. Regardless of whether the CBCA module is added, the performance of our proposed network is better than other networks. Specifially, our network achieved the highest PA (97.34%) and MIoU (94.77%). DeepLabV3+ contains the highest number of parameters and calculations, and the training time is the longest, but its performance is lower than FCN8s, which has the least number of parameters, indicating that overfitting occurs during the training process. Compared with UNet, our network reduces a certain number of parameters and FLOPs, shortening training time while improving performance. After adding the CBCA module to our network, the PA and F_1-Score did not change significantly, but the MIoU improved slightly, indicating that the addition of this module has improved the region smoothness [43]. The above results all verify the effectiveness of our proposed network. Meanwhile, it can be concluded from the number of parameters and FLOPs in the Table 4 that the improvement of our network performance is not simply at the expense of increasing the number of parameters and FLOPs.

Table 4. Results on Green Space testing set. (Bold represents the best result).

Model	PA (%)	F_1-Score	MIoU (%)	Params (M)	FLOPs (G)	Train Time (min)
FCN8S	95.35	94.60	91.49	18.64	20.14	805
Unet	96.60	95.71	94.61	31.04	46.20	795
Deeplab	93.87	92.48	89.79	55.70	82.67	1150
CRUNet	97.33	**96.32**	94.68	28.09	37.46	780
CRAUNet	**97.34**	96.26	**94.77**	28.44	37.46	806

We selected typical urban green space areas to clarify the reasons for the above accuracy differences more clearly. Figure 9 shows several examples of segmentation results of different semantic segmentation networks in different categories of remote sensing images.

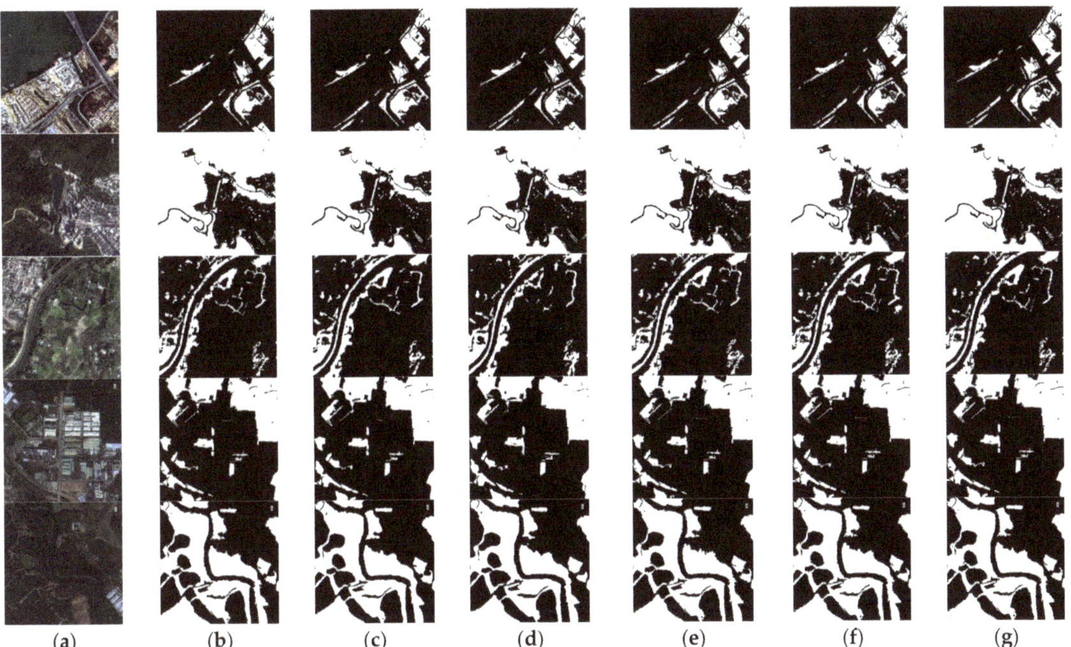

Figure 9. The results classified by the five CNNs on the test set. (**a**) The original images; (**b**) Corresponding labels; (**c**) The predicted maps of FCN8s, (**d**) The predicted maps of DeepLabV3+; (**e**) The predicted maps of UNet; (**f**) The predicted maps of CRUNet; (**g**) The predicted maps of CRAUNet.

It can be seen that whether in urban areas, suburbs or mountains, our network extraction results are closest to the ground truth figure. In the first and second rows, the proposed network removes the influence of shadow and other interference in complex background, indicating that the network has advantages in eliminating noise while other networks are affected by different degrees of noise. In the third and fourth lines, our network extracts green space information more completely. The reason is that the residual module and the channel attention mechanism rely on each other between channels, enhancing the network's feature learning ability. Other CNNs can accurately extract green space but the boundaries extracted by DeeplabV3+ and FCN8s are smoother. To a certain extent, the confusion phenomenon caused similarity or heterogeneity in the remote sensing image is alleviated. Combined with the results of the fifth row, the edges in the prediction results we extracted are more precise, especially at the junction of buildings and green spaces. Compared with the results of the other four pictures, our edges are more detailed. The results show that compared with other networks, our proposed network segmentation results are more complete, and it also has advantages in small target recognition, so it achieved the highest accuracy. To further illustrate the performance of our proposed network for extracting urban green space, Figure 10 shows the example results of CRUNet and CRAUNet on the test set. It can be seen from the figure that the network with or without the CBCA module can identify larger green space targets, but the latter is more susceptible to noise, and there are obvious defects in the identification of smaller feature information. The above results prove that our network has strong feature extraction capabilities and high-resolution detail recovery capabilities for high-resolution remote sensing images. Specifically, the residual structure enhances the feature expression ability of the network and provides richer feature information for the decoder. The CBCA module strengthens the expression of effective features, suppresses invalid features, and provides location guidance for the recovery of high-resolution remote sensing images.

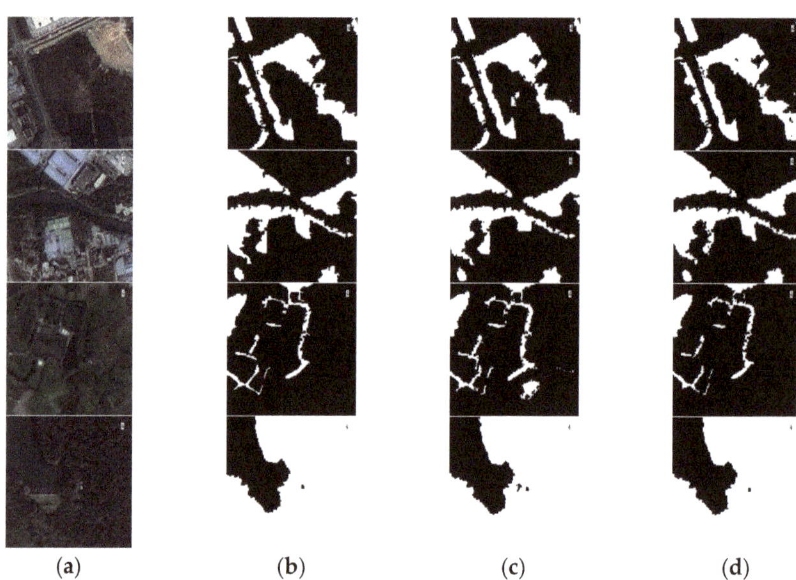

(a)   (b)   (c)   (d)

Figure 10. Results comparison between CRUNet and CRAUNet. (a) The original images; (b) Corresponding labels; (c) The predicted maps of CRUNet; (d)The predicted maps of CRAUNet.

4.2. Performance Comparison on Different Landcover

In order to further evaluate the generality of CRAUNet model, we extracted different types of urban green spaces and easily confused areas from the results, and compared the results extracted from FCN8s, UNet and DeepLabV3+ based on visual inspection and overall accuracy to explain the results. Here, we mainly choose the park green space, residential green space, protective green space, farmland and sports ground and other types of landcovers. The specific features of these landcovers shown in Figure 10 are: forest, golf course, bare land, sports ground, farmland and aquaculture areas

The comparison result is shown in Figure 11 and Table 5. For large areas of green space such as forests and artificial grasses such as golf courses, the four algorithms can accurately identify them. But for bare land with sparse vegetation, FCN8s lost some small patches of green space information. For high-density building areas, the results extracted by FCN8s and DeepLabV3+ algorithms lose a lot of boundary information. And for sports fields containing artificial grass, all algorithms have a certain degree of confusion. But it can be seen from the figure that our network has the lowest error rate. In terms of distinguishing between green space and cultivated land, since more attention was paid to cultivated land during the training process, each algorithm showed better performance. In addition, for built-up areas and aquaculture areas, which are disturbed by shadows and nutrients in the water, compared with our proposed network, the results extracted by other algorithms contain more noise.

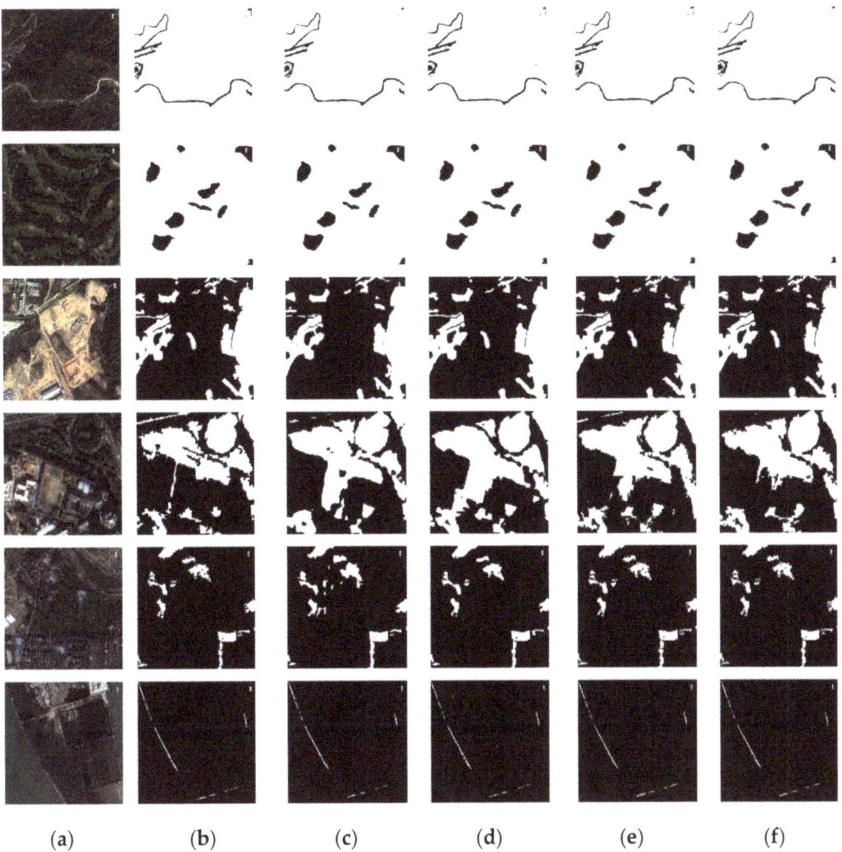

(a)   (b)   (c)   (d)   (e)   (f)

Figure 11. Typical Urban green space classification results. (**a**) The original images; (**b**) corresponding labels; (**c**) the predicted maps of FCN8s, (**d**) the predicted maps of DeepLabV3+; (**e**) the predicted maps of UNet; (**f**) the predicted maps of CRAUNet.

Table 5. The accuracy of different landcover.

Type	FCN8s	DeepLabV3+	UNet	CRAUNet
Froest	96.31%	95.89%	97.42%	97.81%
Golf Course	93.72%	94.34%	98.26%	97.98%
Bare Land	93.68%	95.12%	96.49%	97.06%
Sports Ground	83.91%	80.82%	89.17%	92.33%
Farmland	93.26%	94.05%	96.42%	97.27%
Aquaculture Area	96.32%	95.14%	96.54%	98.36%

From Figure 10 and Table 5 we can see that the performance of CRAUNet is better than that of the other networks, and it achieves high accuracy for each picture. FCN8s and DeepLabV3+ algorithm lost a lot of detailed information of urban green space in the extraction of green space, resulting in that some small green spaces that could not be identified and the boundaries were more blurred. Although UNet can more accurately extract the information of urban green space, it is susceptible to the influence of features with similar spectral characteristics and produces more noise. Our network has shown better performance for extracting different types of green space, and the results are better than other algorithms.

4.3. Urban Green Space Mapping and Accuracy Evaluation

In this section, we used CRAUNet to extract the urban green space of Shenzhen for 2020 based the seven images mentioned in Section 2.1. Compared with common images, remote sensing images have larger data volume. The CNN cannot process such a large image directly. In order to enable CRAUNet to extract the urban green space of a whole remote sensing image, we employed CRAUNet to obtain urban green space information based on a sliding-window method, and the size of a sliding window is 512 × 512 pixels. After extracting urban green space from the seven satellite images, we produced the map of urban green space in Shenzhen with 2 m spatial resolution. The result is shown in Figure 12.

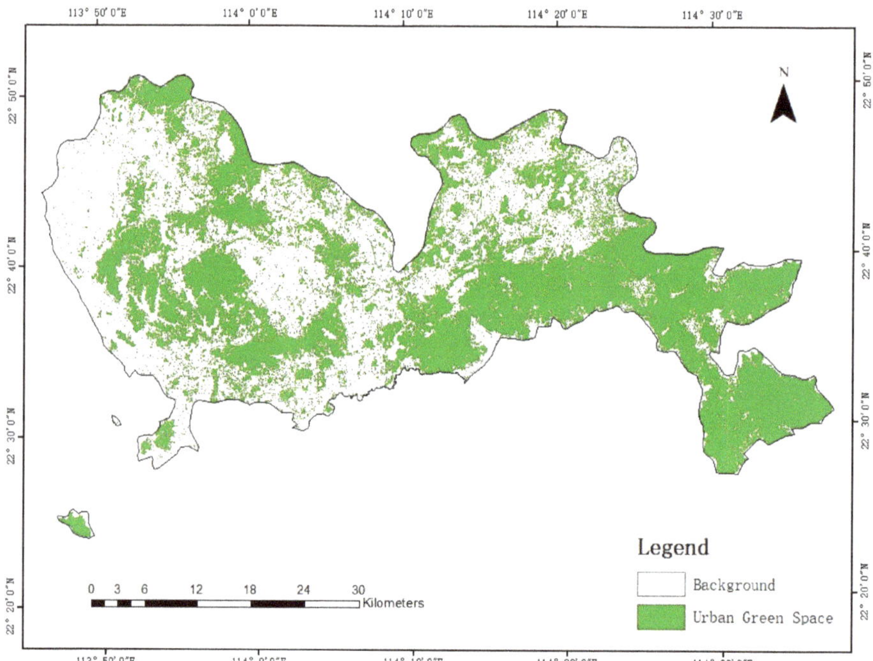

Figure 12. Shenzhen urban green space coverage map in 2020.

In the accuracy evaluation part, we used two kinds of sampling methods to select validation points by ArcGIS and ENVI software. The first method is randomly selecting 500 validation points, and the second method is selecting 491 points for verification at an equal distance. The distribution of sampling points is shown in Figure 13. The confusion matrix and the accuracy evaluation indicators mentioned in Section 2.2.6 were employed to evaluate the mapping accuracy of urban green space extraction. The results are shown in Tables 6 and 7. The accuracy evaluation results of the two sampling methods are different. The accuracy evaluation index of equidistant sampling method is higher than that of random sampling method. Judging from the results in the table, the method has shown high accuracy in the extraction of regional urban green space.

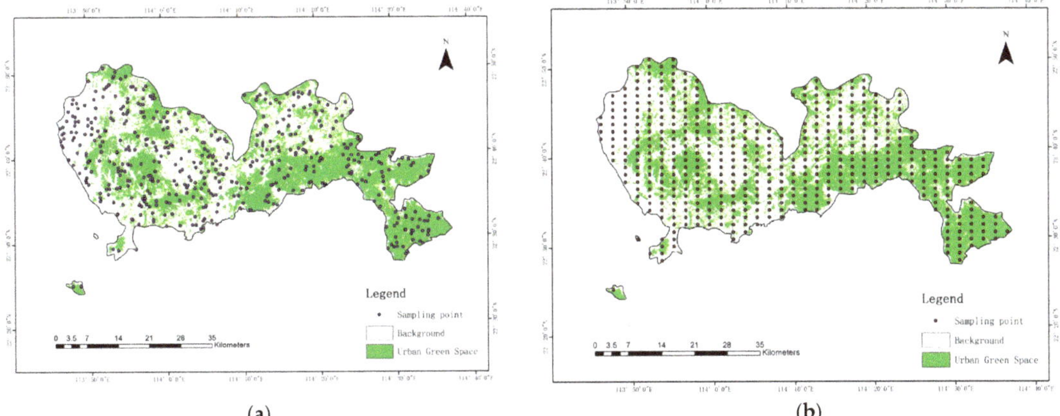

Figure 13. Comparison of different sampling methods for accuracy verification (**a**) Random sampling method; (**b**) Equal distance sampling method.

Table 6. Confusion matrix for urban green space mapping.

Method			Classification Data		
			Urban Green Space	Background	Total
Random sampling	Reference data	Urban Green Space	274	19	293
		Background	14	193	210
		Total	288	212	500
Equal distance sampling		Urban Green Space	258	15	273
		Background	7	211	218
		Total	265	226	491

Table 7. Accuracy assessment for mapping.

Method	PA	MIoU	Precision	Recall	F_1-Score
Random samplings	93.40%	87.32%	93.52%	95.14%	94.32%
Equal distance sampling	95.52%	91.35%	94.51%	97.36%	95.90%

5. Discussion

The high-resolution remote sensing image provides a reliable data support for extracting urban green space accurately. Due to the self-learning ability of deep learning methods, the CNN has been widely used in urban green space extraction. Compared with traditional methods, the CNN can obtain features independently without selecting features artificially. However, the existing network still has some problems in extracting urban green space. In order to solve these problems, we made the following improvements to CRAUNet. According to the analysis in the previous sections, we can conclude that compared with other semantic segmentation algorithms, CRAUNet has the following three advantages in urban green space extraction:

(1) Some networks have made contributions to solving the gradient disappearance phenomenon in the training process, like DenseNet [44]. The core idea of DenseNet is to establish the connection relationship between different layers, make full use of the feature, and further reduce the problem of gradient disappearance., but the price in exchange is a sharp increase in memory usage. In order to make full use of the characteristics of different levels to solve the degradation problem of neural network, and to avoid the increase of memory, we introduce the residual model. In CRAUNet, the residual module we proposed makes it retain more feature information of the

    original image during the feature extraction process, which solves the degradation problem of the neural network to a certain extent and has advantages in identifying small target green spaces. At the same time, this structure does not cause memory usage to rise sharply.

(2)   Existing networks are all affected by noise to varying degrees, and the edges of green spaces are smooth. Especially in DeepLabV3+, it is not sensitive to the boundaries of urban green space. In order to solve this problem, we introduce the attention mechanism. The function of this attention mechanism is to quickly locate important information, highlight important information, suppress unimportant information and obtain information that is not easy to mine. Our network can obtain more complete target and real edge information during segmentation and is less affected by noise. This is because the added CBCA module enhances deep-level feature extraction through the interdependence of modeling channels, strengthens effective green space features and suppresses invalid features.

(3)   In the studies mentioned in the introduction of this paper and other networks that introduce residual structure for innovation, such as R2UNet [45] and ResUNet-a [28], high accuracy is usually obtained by increasing the network depth or adding other modules to increase the number of parameters. The improvement of our network performance is not at the cost of simply increasing the amount of parameters and computation, but by concatenating the features between different convolution kernels to make the network lightweight. That is to say, the network we propose not only improves the accuracy, but also reduces the number of parameters.

In general, CRAUNet has demonstrated powerful feature extraction capabilities and high-resolution details recovery capabilities in t urban green space extraction based on high-resolution remote sensing image. However, CRAUNet still has room for further improvement. In the future, we will do further work on restoring high-resolution detailed information.

6. Conclusions

In this article, we propose a new convolutional neural network CRAUNet for urban green space extraction from GF-1 high-resolution satellite images to better solve the problems in traditional methods. The idea of residual structure and attention mechanism are introduced into the network, and the residual module and CBCA module are proposed to enhance the feature extraction ability while reducing the amount of network parameters and calculations. The results shows that our proposed network achieved the highest performance, the PA and MIoU of our model were 97.34% and 94.77%. Based on the accuracy evaluation result and the visual comparison we can conclude that the performance of CRAUNet is better than that of FCN8s, UNet and DeepLabV3+. In addition, the research in this article is conducive to the use of a large number of high-resolution remote sensing images to dynamically monitor urban green spaces and provide decision-making support for the sustainable development of the urban environment.

Author Contributions: G.M., G.H., G.W. conceived of and designed the experiments. G.W. provided the original remote sensing data. G.M. made the dataset and performed the experiments. G.M. wrote the whole paper. All authors revised this paper and gave some appropriate suggestions. All authors have read and agreed to the published version of the manuscript.

Funding: This research was financially supported by the National Natural Science Foundation of China (61731022), the Strategic Priority Research Program of the Chinese Academy of Sciences (XDA19090300) and the National Key Research and Development Program of China—rapid production method for large-scale global change products (2016YFA0600302).

Institutional Review Board Statement: Not applicable.

Informed Consent Statement: Not applicable.

Data Availability Statement: Not applicable.

Conflicts of Interest: The authors declare no conflict of interest.

References

1. Brander, L.M.; Koetse, M.J. The value of urban open space: Meta-analyses of contingent valuation and hedonic pricing results. *J. Environ. Manag.* **2011**, *92*, 2763–2773. [CrossRef]
2. Kabisch, N.; Strohbach, M.; Haase, D.; Kronenberg, J. Urban green space availability in European cities. *Ecol. Indic.* **2016**, *70*, 586–596. [CrossRef]
3. Völker, S.; Kistemann, T. Developing the urban blue: Comparative health responses to blue and green urban open spaces in Germany. *Health Place* **2015**, *35*, 196–205. [CrossRef] [PubMed]
4. Taylor, L.; Hochuli, D.F. Defining greenspace: Multiple uses across multiple disciplines. *Landsc. Urban Plan.* **2017**, *158*, 25–38. [CrossRef]
5. Standard for Classification of Urban Green Space 2017. CJJ/T 85-2017. Available online: http://www.mohurd.gov.cn/wjfb/2018 06/t20180626_236545.html (accessed on 15 September 2021).
6. Wolch, J.R.; Byrne, J.; Newell, J.P. Urban green space, public health, and environmental justice: The challenge of making cities 'just green enough'. *Landsc. Urban Plan.* **2014**, *125*, 234–244. [CrossRef]
7. Strohbach, M.W.; Arnold, E.; Haase, D. The carbon footprint of urban green space—A life cycle approach. *Landsc. Urban Plan.* **2012**, *104*, 220–229. [CrossRef]
8. Heidt, V.; Neef, M. Benefits of urban green space for improving urban climate. In *Ecology, Planning, and Management of Urban Forests*; Springer: Berlin/Heidelberg, Germany, 2008; pp. 84–96. [CrossRef]
9. Zhang, B.; Gao, J.-X.; Yang, Y. The cooling effect of urban green spaces as a contribution to energy-saving and emission-reduction: A case study in Beijing, China. *Build. Environ.* **2014**, *76*, 37–43. [CrossRef]
10. Fuller, R.A.; Irvine, K.N.; Devine-Wright, P.; Warren, P.H.; Gaston, K.J. Psychological benefits of greenspace increase with biodiversity. *Biol. Lett.* **2007**, *3*, 390–394. [CrossRef] [PubMed]
11. Villeneuve, P.J.; Jerrett, M.; Su, J.G.; Burnett, R.T.; Chen, H.; Wheeler, A.J.; Goldberg, M.S. A cohort study relating urban green space with mortality in Ontario, Canada. *Environ. Res.* **2012**, *115*, 51–58. [CrossRef]
12. Gong, C.; Yu, S.; Joesting, H.; Chen, J. Determining socioeconomic drivers of urban forest fragmentation with historical remote sensing images. *Landsc. Urban Plan.* **2013**, *117*, 57–65. [CrossRef]
13. Wang, H.; Wang, C.; Wu, H. Using GF-2 imagery and the conditional random field model for urban forest cover mapping. *Remote Sens. Lett.* **2016**, *7*, 378–387. [CrossRef]
14. Zhou, X.; Li, L.; Chen, L.; Liu, Y.; Cui, Y.; Zhang, Y.; Zhang, T. Discriminating urban forest types from Sentinel-2A image data through linear spectral mixture analysis: A case study of Xuzhou, East China. *Forests* **2019**, *10*, 478. [CrossRef]
15. Yao, Z.; Liu, J.; Zhao, X.; Long, D.; Wang, L. Spatial dynamics of aboveground carbon stock in urban green space: A case study of Xi'an, China. *J. Arid Land* **2015**, *7*, 350–360. [CrossRef]
16. Myeong, S.; Nowak, D.J.; Duggin, M.J. A temporal analysis of urban forest carbon storage using remote sensing. *Remote Sens. Environ.* **2006**, *101*, 277–282. [CrossRef]
17. Shojanoori, R.; Shafri, H.Z. Review on the use of remote sensing for urban forest monitoring. *Arboric Urban* **2016**, *42*, 400–417. [CrossRef]
18. Shen, C.; Li, M.; Li, F.; Chen, J.; Lu, Y. Study on urban green space extraction from QUICKBIRD imagery based on decision tree. In Proceedings of the 2010 18th International Conference on Geoinformatics, Beijing, China, 18–20 June 2010; pp. 1–4. [CrossRef]
19. Feng, Q.; Liu, J.; Gong, J. UAV remote sensing for urban vegetation mapping using random forest and texture analysis. *Remote Sens.* **2015**, *7*, 1074–1094. [CrossRef]
20. Diamantopoulou, M.J. Filling gaps in diameter measurements on standing tree boles in the urban forest of Thessaloniki, Greece. *Environ. Model. Softw.* **2010**, *25*, 1857–1865. [CrossRef]
21. Walker, J.S.; Briggs, J.M. An object-oriented approach to urban forest mapping in Phoenix. *Photogramm. Eng. Remote Sens.* **2007**, *73*, 577–583. [CrossRef]
22. Ardila, J.P.; Bijker, W.; Tolpekin, V.A.; Stein, A. Context-sensitive extraction of tree crown objects in urban areas using VHR satellite images. *Int. J. Appl. Earth Obs. Geoinf.* **2012**, *15*, 57–69. [CrossRef]
23. Nijhawan, R.; Sharma, H.; Sahni, H.; Batra, A. A deep learning hybrid CNN framework approach for vegetation cover mapping using deep features. In Proceedings of the 2017 13th International Conference on Signal-Image Technology & Internet-Based Systems (SITIS), Jaipur, India, 4–7 December 2017; pp. 192–196. [CrossRef]
24. A Novel Software for Analyzing the Natural Landscape Surface Using Deep Learning. Available online: https://openreview.net/forum?id=HJgWtnd9kr (accessed on 15 September 2021).
25. Timilsina, S.; Aryal, J.; Kirkpatrick, J.B. Mapping urban tree cover changes using object-based convolution neural network (OB-CNN). *Remote Sens.* **2020**, *12*, 3017. [CrossRef]
26. Hartling, S.; Sagan, V.; Sidike, P.; Maimaitijiang, M.; Carron, J. Urban tree species classification using a WorldView-2/3 and LiDAR data fusion approach and deep learning. *Sensors* **2019**, *19*, 1284. [CrossRef]

27. Blaschke, T.; Hay, G.J.; Kelly, M.; Lang, S.; Hofmann, P.; Addink, E.; Feitosa, R.Q.; Van der Meer, F.; Van der Werff, H.; Van Coillie, F. Geographic object-based image analysis–towards a new paradigm. *ISPRS J. Photogramm. Remote Sens.* **2014**, *87*, 180–191. [CrossRef] [PubMed]

28. Diakogiannis, F.I.; Waldner, F.; Caccetta, P.; Wu, C. ResUNet-a: A deep learning framework for semantic segmentation of remotely sensed data. *ISPRS J. Photogramm. Remote Sens.* **2020**, *162*, 94–114. [CrossRef]

29. Kemker, R.; Salvaggio, C.; Kanan, C. Algorithms for semantic segmentation of multispectral remote sensing imagery using deep learning. *ISPRS J. Photogramm. Remote Sens.* **2018**, *145*, 60–77. [CrossRef]

30. Albawi, S.; Mohammed, T.A.; Al-Zawi, S. Understanding of a convolutional neural network. In Proceedings of the 2017 International Conference on Engineering and Technology (ICET), Antalya, Turkey, 21–23 August 2017; pp. 1–6. [CrossRef]

31. Yuan, Y.; Guo, H.; Xu, H.; Li, W.; Luo, S.; Lin, H.; Yuan, Y. China's first special economic zone: The case of Shenzhen. In *Building Engines for Growth and Competitiveness in China: Experience with Special Economic Zones and Industrial Clusters*; Zeng, D.Z., Ed.; World Bank Publications: Washington, DC, USA, 2010; Chapter 2; pp. 55–86.

32. Chen, X.; Li, F.; Li, X.; Hu, Y.; Hu, P. Evaluating and mapping water supply and demand for sustainable urban ecosystem management in Shenzhen, China. *J. Clean. Prod.* **2020**, *251*, 119754. [CrossRef]

33. Tengfei, L.; Weili, J.; Guojin, H. Nested regression based optimal selection (NRBOS) of rational polynomial coefficients. *Photogramm. Eng. Remote Sens.* **2014**, *80*, 261–269. [CrossRef]

34. Xiao, P.; Zhang, X.; Zhang, H.; Hu, R.; Feng, X. Multiscale optimized segmentation of urban green cover in high resolution remote sensing image. *Remote Sens.* **2018**, *10*, 1813. [CrossRef]

35. Yang, G.; Pennington, J.; Rao, V.; Sohl-Dickstein, J.; Schoenholz, S.S. A mean field theory of batch normalization. *Arxiv* **2019**, arXiv:1902.08129 2019.

36. He, K.; Zhang, X.; Ren, S.; Sun, J. Deep residual learning for image recognition. In Proceedings of the IEEE Conference on Computer Vision and Pattern Recognition, Las Vegas, NV, USA, 26 June–1 July 2016; pp. 770–778.

37. He, K.; Zhang, X.; Ren, S.; Sun, J. Identity mappings in deep residual networks. In Proceedings of the European Conference on Computer Vision, Amsterdam, The Netherlands, 11–14 October 2016; pp. 630–645. [CrossRef]

38. Wang, F.; Jiang, M.; Qian, C.; Yang, S.; Li, C.; Zhang, H.; Wang, X.; Tang, X. Residual attention network for image classification. In Proceedings of the IEEE Conference on Computer Vision and Pattern Recognition, Honolulu, HI, USA, 21–26 July 2017; pp. 3156–3164.

39. Woo, S.; Park, J.; Lee, J.-Y.; Kweon, I.S. Cbam: Convolutional block attention module. In Proceedings of the European Conference on Computer Vision (ECCV), Munich, Germany, 8–14 September 2018; pp. 3–19.

40. Ronneberger, O.; Fischer, P.; Brox, T. U-net: Convolutional networks for biomedical image segmentation. In Proceedings of the International Conference on Medical Image Computing and Computer-Assisted Intervention, Munich, Germany, 5–9 October 2015; pp. 234–241. [CrossRef]

41. Odena, A.; Dumoulin, V.; Olah, C. Deconvolution and checkerboard artifacts. *Distill* **2016**, *1*, e3. [CrossRef]

42. Yu, L.; Wu, H.; Zhong, Z.; Zheng, L.; Deng, Q.; Hu, H. TWC-Net: A SAR ship detection using two-way convolution and multiscale feature mapping. *Remote Sens.* **2021**, *13*, 2558. [CrossRef]

43. Badrinarayanan, V.; Kendall, A.; Cipolla, R. Segnet: A deep convolutional encoder-decoder architecture for image segmentation. *IEEE Trans. Pattern Anal. Mach. Intell.* **2017**, *39*, 2481–2495. [CrossRef] [PubMed]

44. Gao, H.; Zhuang, L.; Van Der Maaten, L.; Kilian, Q.W. Densely Connected Convolutional Networks. In Proceedings of the IEEE Conference on Computer Vision and Pattern Recognition (CVPR), Honolulu, HI, USA, 21–26 July 2017; pp. 4700–4708.

45. Alom, M.Z.; Yakopcic, C.; Hasan, M.; Taha, T.; Asari, V. Recurrent residual U-Net for medical image segmentation. *J. Med Imaging* **2019**, *6*, 014006. [CrossRef] [PubMed]

Article

Responses of Visual Forms to Neighboring Competition: A Structural Equation Model for *Cotinus coggygria* var. *cinerea* Engl.

Yujuan Cao [1], Jiyou Zhu [1], Chengyang Xu [1,*] and Richard J. Hauer [2]

[1]   Research Center for Urban Forestry, Key Laboratory for Silviculture and Forest Ecosystem of State Forestry and Grassland Administration, Key Laboratory for Silviculture and Conservation of Ministry of Education, Beijing Forestry University, Beijing 100083, China; yujuancao@foxmail.com (Y.C.); zhujiyou007@163.com (J.Z.)

[2]   College of Natural Resources, University of Wisconsin-Stevens Point, Stevens Point, WI 54481, USA; rhauer@uwsp.edu

*   Correspondence: cyxu@bjfu.edu.cn; Tel.: +86-010-6233-7082

Abstract: Background: The visual forms of individual trees in peri-urban forests are driven by a complex array of simultaneous cause-and-effect relationships. Materials and Methods: Structural Equation Modeling (SEM), as a specialized analytical technique, was used to model and understand the complex interactions. It was applied to find out responses of visual forms to neighboring competition in a peri-urban forest dominated by *Cotinus coggygria* var. *cinerea* Engl. in Beijing, China. Research Highlights: Light interception and space extrusion have substantial effects on visual forms, expressed as crown forms and foliage forms. The structural model in SEM tested hypothetical correlations among latent variables, namely neighboring competition, crown forms, and foliage forms. Results: The fitted model suggested a direct negative effect of neighboring competition on crown forms and an insignificant negative direct effect on foliage forms. Moreover, an indirect positive effect on foliage forms mediated by crown forms was revealed. Conclusions: The fitted SEM and associated findings should facilitate peri-urban forest landscape management by providing insight into causal mechanisms of visual forms of individual trees and thereby assisting in the visual quality promotion.

Keywords: aesthetics; causal effects; crown form; foliage form; urban forests

Citation: Cao, Y.; Zhu, J.; Xu, C.; Hauer, R.J. Responses of Visual Forms to Neighboring Competition: A Structural Equation Model for *Cotinus coggygria* var. *cinerea* Engl. *Forests* **2021**, *12*, 1073. https://doi.org/10.3390/f12081073

Academic Editors: Tina Harrison and Manuel Esperon-Rodriguez

Received: 5 July 2021
Accepted: 9 August 2021
Published: 11 August 2021

Publisher's Note: MDPI stays neutral with regard to jurisdictional claims in published maps and institutional affiliations.

1. Introduction

The peri-urban forest provides a way for people to connect with nature, and visually attractive landscapes make places more enjoyable for outdoor leisure and entertainment [1,2]. Viewing natural plants is linked to human health improvement and stress reduction [3].

Relationships between tree form and aesthetic preference exist. Varied tree forms effectively improve the attractiveness of forest landscapes and enhance tourist experiences [4]. However, researchers concentrated on growth forms in most of the studies about tree forms. Visual forms of trees were less studied because its complication in describing and representing. In this paper, visual forms were used to represent forms of objects that can be discerned under light [5–7] and represent the visual aspect of tree forms. Generally, visual forms of trees are determined by genes and adaptation to environmental conditions [8]. Moreover, the diversity of visual forms of the same tree species is affected by many factors such as slope, stand density, genetics, and environmental conditions [9,10]. Besides the influence of stand density and species diversity, it is well known that the size-asymmetric competition by neighboring plants also affects visual forms of crown and trunk [11,12].

Despite the widely recognized negative trends in growth-competition relationships in forests [13–16], it remains unclear how visual forms respond to neighboring competition. The processes to assess visual forms' response to neighboring competition involve complicated interactions and many causal relationships. Standard statistical methods, such as analysis of variance [17] (pp. 19–46) and multiple regression [18], may not be enough to follow the mechanisms linking visual forms to the living conditions of trees. A structural equation model (SEM) offers an alternative way to test complicated interactions [19]. Timilsina et al. [20] used SEM to analyze the causal factors of carbon stores of subtropical plants, which used biometric and ecological relationships as part of the model.

In this study, *Cotinus coggygria* var. *cinerea* Engl. was treated as the subject tree to investigate the role of neighboring competitions on visual forms of the individual trees in forests. The neighboring competition was described and evaluated in the competitive unit, containing one subject tree and four neighboring trees. Moreover, visual forms described by several traits in core visual parts of trees and trends in relationships between every trait and competition indicators were focused on. Using SEM, 218 competitive units in the *Cotinus coggygria* forest were set and analyzed to answer the following questions: (i) How do neighboring competitions affect visual forms? (ii) How do crown forms and foliage forms respond to neighboring competition, respectively? (iii) What peri-urban forest managers could do to promote the visual quality of tree forms.

2. Materials and Methodology

2.1. Study Area and Data Collection

The study was carried out in a forest dominated by *Cotinus coggygria* var. *cinerea* Engl. in Badaling National Forest Park, Yanqing District, Beijing, China (40°20′46.01″ N, 116°0′52.20″ E). The park encompasses 2933.3 hectares with a 780-m average elevation and experiences a continental monsoon climate zone with semi-humid and warm temperate climates. A mean 10.8 °C annual average temperature occurs with a 160-days frost-free period and 454 mm average annual precipitation mainly concentrated from July to August. The park was the first ecological public welfare forest passed FSC (Forest Stewardship Council) Certification, and there are 539 species of wild vascular plants and 158 species of wild vertebrates in it. Ninety-six percent of the land in the park is covered by trees and plants, which consist of pure *Cotinus coggygria* forests, broad-leaved mixed forests of *Cotinus coggygria* with *Armeniaca sibirica*, and coniferous and broad-leaved mixed forests of *Cotinus coggygria* with *Platycladus orientalis*. The stand density is between 1212 trees·ha^{-1} and 2688 trees·ha^{-1}, the average canopy area is between 0.17 m^2 and 58.26 m^2, and the average tree height is between 0.7 m and 10.2 m.

In the park, seven 50 m × 50 m sample plots (0.25 ha) of broad-leaved mixed forest of *Cotinus coggygria* were set with poles and measuring ropes aided by compass. A coordinate system was set in each plot with the origin at the corner with magnetic azimuth 215° of the plot. The X-axis was oriented to magnetic azimuth 90°, and the Y-axis was oriented to magnetic azimuth 0°. Besides, the vertical lines where $\chi = 0$, $\chi = 10$, $\chi = 20$, $\chi = 30$, $\chi = 40$, and $\chi = 50$ and the horizontal lines where y = 0, y = 10, y = 20, y = 30, y = 40, and y = 50 were marked out by measuring ropes. Then, the *Cotinus coggygria* (stem collar diameter > 2 cm), whose trunk was closest to the intercept point of two measuring ropes, was marked as the subject tree. When all subject trees in a plot were marked, remove all measuring ropes in the plot. The competitive units were set using the point-centered quarter method [21]. The subject tree was treated as the center of the competitive unit, around which four 90° quarters were set clockwise from magnetic azimuth 0° [22] (pp. 86–87). In each quarter, the tree (stem collar diameter > 2 cm) in any species whose trunk was closest to the trunk of the subject tree was chosen as a neighboring tree. In total, 218 competitive units were chosen in the park. All the competitive units were indexed from 1 to 218.

Data were collected as follows. For all five trees in each competitive unit, stem collar diameter was measured at 0.1 m above ground with a diameter tape. Height to the lowest live branch was measured from the ground with a tape. Tree height was measured with a laser altimeter. Crown width was measured with a tape parallel to ground from the trunk to the most distant point in four directions (east, west, south, and north). Additionally, distances between trunks of neighboring trees and the trunk of the subject tree were measured with tape. Moreover, current year twig and leaves were sampled from the upper, middle, and lower layers of the crown of the subject tree at a random direction according to the particular order. If the index of a competitive unit was divided by 4 with a reminder of 0, the north crown was selected, and 1 for the east, 2 for the south, 3 for the west. The base diameter of the current year twig was measured with an electronic vernier caliper. The fresh and dry weight of leaves were measured with an electronic balance. A scanner (EPSON V39) and *Image J* (National Institutes of Health, https://imagej.nih.gov/ij/ accessed on 10 August 2019) software were employed to quantify the area of leaves after removing petioles. Moreover, fresh leaves were killed at 80 °C in a drying baker and dried to constant weight at 105 °C, and the fresh and dry leaves were weighed with an electronic balance. The measurement accuracy and named abbreviation of measured traits are shown in Table 1.

Table 1. Named abbreviation (Abbr.) and measurement accuracy of measured traits.

Traits	Abbr.	Accuracy	Traits	Abbr.	Accuracy
Base diameter of current year twig	BD	0.02 mm	Height to the lowest live branch	HL	0.01 m
Crown width	CW	0.1 m	Leaf area	LA	0.1 mm^2
Distance	dist	0.01 m	Stem collar diameter	CD	0.1 cm
Dry weight of leaves	DW	0.01 g	Tree height	H	0.1 m
Fresh weight of leaves	FW	0.01 g			

2.2. Overview of Our Approach

To describe visual forms of the subject tree, six primary traits of crown, leaves, and twigs were selected. The definition of visual forms was suggested as a unified whole to present the visual aspects of tree forms (see Section 2.3). *Cotinus coggygria*, as a multiple-branching deciduous tree with an open, spreading, irregular crown, was treated as subject trees. It is more complicated to estimate and evaluate the competition it suffered from neighboring trees than a single trunk deciduous tree with a broad, rounded crown. Therefore, we redefined competition caused by neighboring trees referring to the proposed competition indexes (see Section 2.4). To determine the relevancy of visual traits and neighboring competition, each competition indicator was divided into four continuous disjoint intervals by their quartiles (0–25%, 25–50%, 50–75%, 75–100%). Furthermore, all selected traits in every interval were summarized, respectively).

A structural equation model (SEM) was used for testing hypothetical relationships among all observed variables [19,23] in this study. As the application of SEM requires a set of well-defined hypotheses generated from theoretical considerations, previous knowledge, and personal observation [24], three hypotheses about neighboring competition and visual forms (see Section 2.5) were proposed. We transformed complex interactions into directional path networks for linking observed variables and then constructed latent variables and the structural model based on proposed hypotheses (see Section 2.6).

The SEM was evaluated from multivariate data collected in 218 competitive units. Moreover, some basic fit statistics were selected from the comprehensive listing of indexes and criteria provided by Kline [25] and Schumacker and Lomax [23] to assess model fit. Furthermore, results that indicate the cause-and-effect relationships were analyzed with the proposed hypotheses.

2.3. Visual Forms of the Individual Tree

In visual arts, form refers to three-dimensional objects with a volume of height, width, and depth [26]. Visual forms in this paper represent the visual aspects of individual trees, which were considered the three-dimensional physical entity with aesthetic value. Visual forms used in the study idealized some structural complexity of tree shapes, which referred to the reciprocal arrangement between the different organs. For example, the geometry and topology of the tree crown were abstracted as a simple circle.

The height of a dominant tree species and the shape of its foliage determine the overall landscape of many terrestrial plant communities [27]. This paper described the visual forms of a tree as two different systematic units, namely crown forms and foliage forms. Owing to the multiple-branch features, the trunks of *Cotinus coggygria* are generally short, so crown forms in this study were represented as the combination of trunk form and crown form. Furthermore, the visual forms of an individual tree were roughly divided into four categories according to the value of crown form and foliage form. In Figure 1, we see four different categories of visual forms laid out along two dimensions. The dimension on the horizontal axis is concerned with crown forms, whereas the ones on the vertical axis address foliage forms. The foliage forms of trees in the categories on top are dense, while the foliage forms in the bottom categories are sparse. The trees in categories on the left are with a small crown, and the ones on the right are with a large crown.

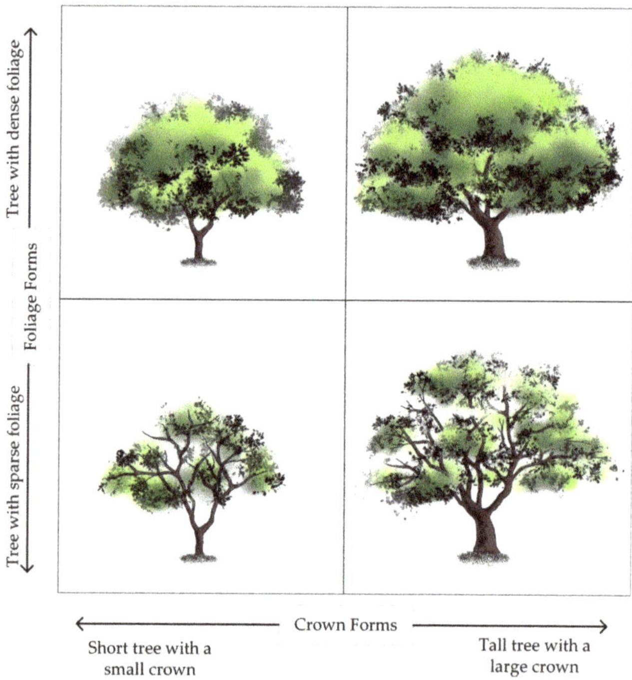

Figure 1. Some visual forms of an individual tree, organized into four categories. Four different visual forms were laid out along two dimensions, referred to crown forms and foliage forms.

In this study, three traits, namely tree height, crown width, and crown length, were selected from collected data to depict crown forms. Moreover, dry weight and total area of leaves on current year twigs and the base diameter of current year twigs were another three primary traits selected to represent foliage forms. Table 2 presents descriptive statistics of observed variables used to depict crown forms and foliage forms.

Table 2. Descriptive statistics of visual form traits ($n = 218$).

Construct	Traits	Unit	Min.	Max.	Average	Standard Deviation
Crown Forms	Crown Length	m	1.08	4.49	2.94	0.87
	Crown Width	m	1.07	6.75	3.62	1.17
	Tree Height	m	1.68	6.41	3.58	0.91
Foliage Forms	Diameter of Current Year Twigs	mm	1.33	3.68	2.31	0.48
	Dry Weight of Leaves	g	0.03	0.64	0.19	0.13
	Leaves Area	mm^2	3501	28,588	13,882	4810

2.4. Neighboring Competition

Neighboring competition in this paper referred to that trees in a competitive unit compete for growing space and sunlight. The growth states and visual forms of subject trees were directly linked to their capability to obtain growing space and presumably resource uptake [28,29]. Moreover, the crown structure of a subject tree could be affected by neighboring trees that occupy growing space [30] and absorb sunlight [31].

Trees in a competitive unit share growing space. Bella [32] proposed a model, competitive influence zone overlap, to describe and evaluate inter-tree competitions. Moreover, the influence zone was defined as an area over which the tree obtains or competes for site factors [33]. In the study, competitive influence zone overlap was used as a measurable surrogate for space competition. Figure 2 shows the influence zone overlap between trees in a competitive unit. The maximum zone of influence of a species is related to its open-grown crown size [32]. Tree crowns were simplified as circles with a radius calculated from the average crown width in four directions [34]. The space competition (*SC*) could be calculated in formula 1, in which AO_i, the overlapping area of the ith neighboring tree, could be calculated by the law of cosines and sector area formula. The A_0 refers to the crown projected area of the subject tree.

$$SC = \sum_{i=1}^{4} \frac{AO_i}{A_0} \tag{1}$$

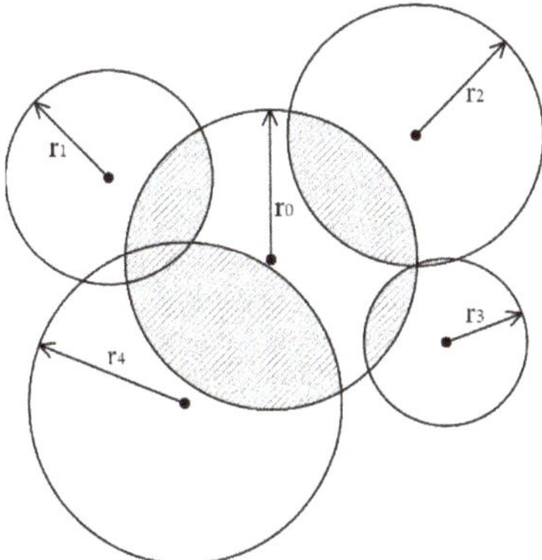

Figure 2. Overlap area of subject tree (the dark part). The crown width of subject tree was r$_0$, crown widths of neighboring trees were r$_1$, r$_2$, r$_3$, and r$_4$.

Each tree attempts to maximize the net productivity under the light environment, which is determined by neighboring trees [27]. Trees with a higher crown than their neighbors can avoid shading by competitors and will enjoy likely be more productive. More accurately, sunlight intensity would increase with relative tree height, as a relatively tall tree with taller neighbors would still be shaded. In the broad-leaved forest, crown shyness, which describes the phenomenon whereby tree crowns avoid growing into each other, producing a puzzle-like pattern of complementary tree crowns in the canopy [35], was found between individual trees [36]. As competitions lead to trees growing asymmetrical crowns instead of their ideal symmetrical shape, measurements related to tree size rather than that representative of the ideal symmetrical shape were chosen. Therefore, the stem collar diameter of neighboring trees (CD_i) and distances ($dist_i$) between neighboring trees and the subject tree should be considered. Sunlight competition (LC) in Formula 2 from Smith and Bell [37].

$$LC = \sum_{i=1}^{4} \frac{H_0 \times CD_i}{dist_i \times CD_0} \tag{2}$$

When two trees of different sizes compete in a forest stand, they do not affect each other equally. To maintain a higher growth rate with a generally lower metabolic efficiency, the larger tree must exploit the site considerably beyond what is proportional to its size [38]. In other words, trees with larger biomass usually dominate the competitive units [39]. Moreover, the smaller tree with a lower growth rate and higher metabolic efficiency can exist on fewer resources than its proportional share. Thus, the growth state of trees should also be considered in the description and evaluation of inter-tree competition. Relative competitiveness (RC) was defined to depict the relative pressure subject trees suffered to obtain living resources in a competitive unit. The relative competitiveness could be calculated in Formula 3, where CW_0 refers to the crown width of the subject tree, and the CW_i refers to the crown width of the ith neighboring tree. Table 3 presents descriptive statistics of three neighboring competition indicators used in the study.

$$RC = \frac{1}{4} \sum_{i=1}^{4} \frac{H_i \times CW_i^2}{H_0 \times CW_0 \times dist_i} \tag{3}$$

Table 3. Descriptive statistics of neighboring competition indicators ($n = 218$).

Neighboring Competition Indicator	Min.	Max.	Average	Standard Deviation
Relative Competitiveness	0.02	2.68	0.55	0.48
Space Competition	0.00	2.46	0.46	0.47
Sunlight Competition	1.44	35.36	7.79	5.39

2.5. Hypotheses

According to theoretical considerations, previous knowledge, and personal observation, three hypotheses about neighboring competition and visual forms of trees were proposed.

Hypothesis 1 (H1). *Neighboring competition directly affects crown forms.*

Trees growing under different light conditions have very different crown shapes, which has been explained as an optimal strategy [40]. For example, the crown of an isolated oak tree is of a hemispherical shape, in which the amount of trunk and branches appear to be the minimum necessary to keep the leaves apart from each other. In contrast, a dominant oak of the same species growing in a dense forest has a long trunk, and the foliage is more so at the top part of the trunk. Additionally, a tree with a long trunk and a higher crown than its neighbors can avoid shading by competitors and will likely result in greater photosynthetic productivity [27]. A similar hypothesis could be suggested in the forest dominated by *Cotinus coggygria*.

Hypothesis 2 (H2). *Neighboring competition affects foliage forms directly and indirectly mediated by crown forms.*

Trees compete with each other and with other plants for the sunlight available on a site. When trees get over-topped and shaded by others, their access to sunlight is reduced or eliminated. The leaves on shaded conduct reduced photosynthesis but still cost the plant energy, nutrients, and water to maintain. These branches generally die and may self-prune after they can no longer maintain themselves. Therefore, it is hypothesized that neighboring competition directly affects foliage forms.

Plant growth was a process in which leaves grow, and then more light was captured through photosynthesis to produce and continue the plant growth as estimated by dry mass [41]. The total leaf area should be coordinated with stem diameter for both mechanical [42] and hydraulic reasons [43]. Moreover, the higher part of the foliage shades the lower part, but the lower does not affect the higher. So, foliage forms were also affected by crown forms.

Hypothesis 3 (H3). *Crown forms and foliage forms respond differently to neighboring competition.*

Competitive interactions between plants were mediated by their growth states [44]. Different parts of trees respond differently to neighboring competition. In uncrowded groups, tree height has a linear relation with total leaf area or branch length, but this relation becomes curvilinear or discontinuous in overcrowded groups [44]. Furthermore, interspecific leaf size variation includes space pressures and environment adaptations associated with optimizing photosynthesis. It could be inferred that neighboring competition affects crown forms and foliage forms differently.

2.6. SEM and Measurement Models

Generally, SEM presents two components, a measurement model for evaluating unobserved latent variables or linear function structures of observed variables and a structural model indicating the direction and strength of the relationship of latent variables [45] (pp. 141–152).

In Figure 3, ovals and rectangles represented latent variables and observed variables, respectively. Unobserved latent variables were evaluated as linear function structures of observed variables. The neighboring competition was treated as a multifaceted concept with three observed variables measuring its different facets. Space competition, sunlight competition, and relative competitiveness were used as reflective indicators to construct the latent variable named neighboring competition. Tree height, crown width, and crown length were used as reflective indicators to construct the latent variable named crown forms. The dry weight of leaves, area of leaves, and base diameter of current year twigs were used as reflective indicators to construct the latent variable named foliage forms.

Analysis of a Moments Structure (*AMOS*) version 20 (IBM, Chicago, IL, USA) was used to estimate path coefficients and test effects among variables in the SEM. Confirmatory factor analysis was used to evaluate measurement models. The construct reliability (CR) was evaluated with Cronbach's alpha. Discriminant validity was evaluated with average variance extracted (AVE) [23,25].

Regression coefficients of each SEM path were estimated with the maximum likelihood method, and four basic fit statistics were chosen to assess model fit are as follows. A chi-squared (χ^2) difference between observed and estimated covariance matrices was used to test the significance of the overall model fit. The Goodness of Fit Index (GFI) measures the relative amount of variance and covariance jointly accounted for by the model, and GFI ranges from 0 to 1.0, with 1.0 indicating the best fit. Root Mean Square Error of Approximation (RMSEA) measures the degree of misspecification per model degree of freedom, adjusted for sample size. Browne and Cudeck [46] suggested that RMSEA ≤ 0.05 indicated a close approximation or fit, a value between 0.05 and 0.08 indicated a reasonable approximation, and a value ≥ 0.1 suggested a poor fit.

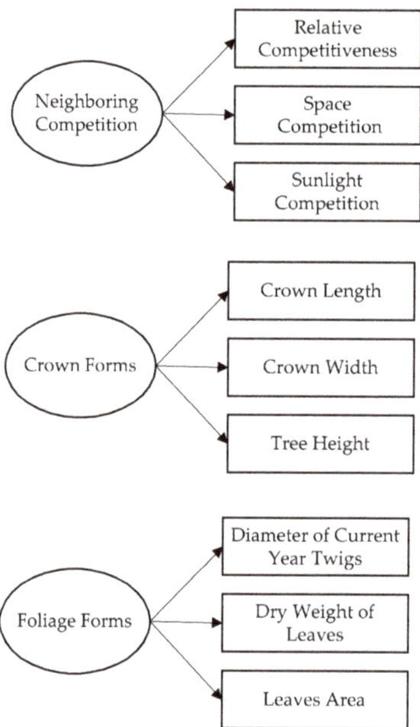

Figure 3. The measurement model used in this study with the ovals depicting a latent variable and the rectangles depict observed variables. Single arrowheads from the latent variable to the observed variables represent direct effects.

3. Results

In this paper, six primary observed traits were used as reflective indicators to construct latent variables that illustrate visual forms of individual trees. We constructed the neighboring competition from three different facets that differ in living resources for which trees competed. Moreover, 218 competitive units covering most of the growth stage of subject trees were set in total.

3.1. Bivariate Relationships among Visual Forms Traits and Competition Indicators

The three competition indicators (space competition, sunlight competition, and relative competitiveness) influence crown forms quite intensely but differing in their intensity for different traits. Increasing competition intensity results in decreasing crown forms trait (Figure 4a). The high potential relativity of crown forms traits and relative competitiveness was found ($R^2 > 0.18$, Figure 5). The effects of space competition on crown form traits are less pronounced than the sunlight competition and relative competitiveness (Figures 4a and 5).

The competition indicators hardly influence foliage form traits. The medians of leaves area and twig diameter remain relatively constant in different intervals of competition indicators (Figure 4b). The median of the dry weight of leaves varies a bit. Especially in the third interval of space competition and sunlight competition, medians of the dry weight of leaves were less than that in the other three intervals. Slight correlations between competition indicators and foliage form traits ($R^2 < 0.03$, Figure 5) were found.

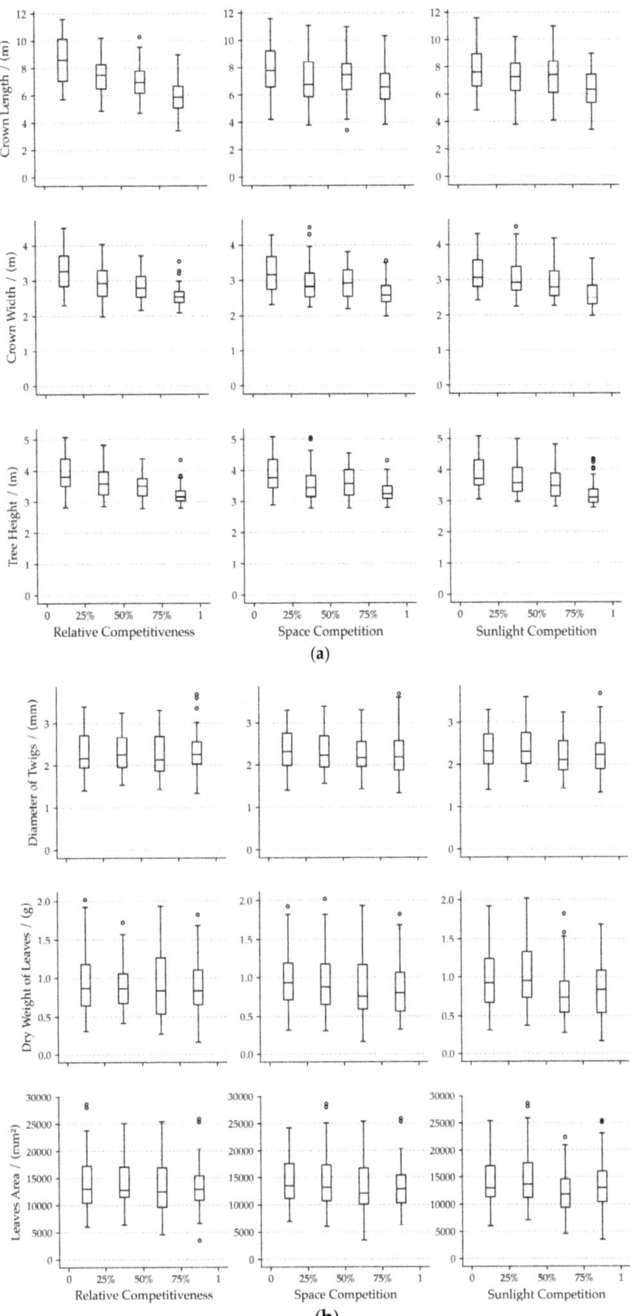

Figure 4. Box-plots of crown form traits (**a**) and foliage form traits (**b**) for each neighboring competition indicator. In the subplots, 25%, 50%, and 75% indicate the lower quartile, median, and upper quartile of each neighboring competition indicator.

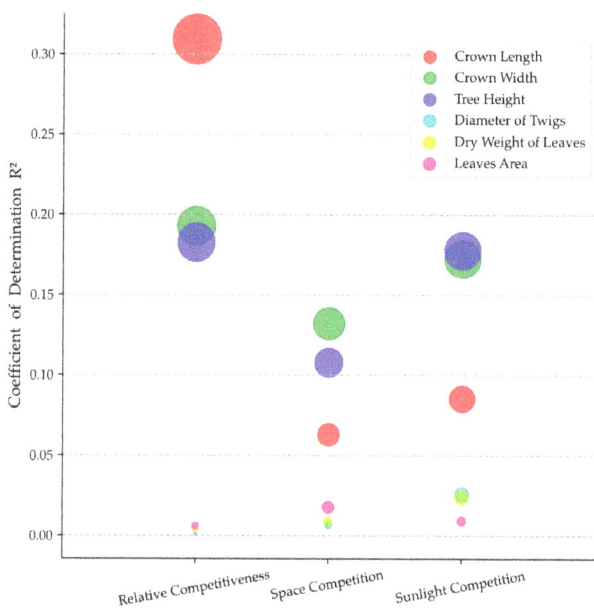

Figure 5. Bubble plot of coefficient of determination (R^2) in linear regression models of each visual form traits and neighboring competition indicators. The size of the bubble was positively related to R^2.

3.2. Confirmatory Factor Analysis of SEM

Confirmatory factor analysis results were shown in Table 4. All observed variables measured the constructs well as determined by the Cronbach's alpha > 0.7 (p-value < 0.05), CR > 0.7 (p-value < 0.05), and AVE > 0.5 (p-value < 0.05). Thus, it was not necessary to delete or modify any variables. The latent variables loaded well on their constructs and measurement models (Figure 3) were sufficient to test the path coefficients [47].

Table 4. Results of confirmatory factor analysis for the reliability and validity of constructs.

Construct	Item	Unstandardized Factor Loading	Cronbach's Alpha	Construct Reliability (CR)	Average Variance Extracted (AVE)
Neighboring Competition	Relative Competitiveness	1.0	0.71	0.80	0.51
	Space Competition	0.88			
	Sunlight Competition	0.66			
Crown Forms	Crown Length	0.65	0.82	0.86	0.68
	Crown Width	0.87			
	Tree Height	1.0			
Foliage Forms	Diameter of Current Year Twigs	0.82	0.76	0.81	0.51
	Dry Weight of Leaves	1.0			
	Leaves Area	0.83			

The model converged with indexes indicating a good overall model fit. The chi-square corrected by the degrees of freedom (χ^2/df) was 1.68 (the χ^2 = 40.2 with df = 23). The GFI was 0.960, and CFI was 0.990, with 1.0 indicating the best fit. Moreover, the RMSEA was 0.059, indicating a reasonable approximation (Table 5).

Table 5. Overall fitness of the proposed model and recommended values (n = 218).

Index	Value	Recommended Value
χ^2/df	1.68	<5.00
Comparative Fit Index (CFI)	0.99	>0.90
Goodness of Fit Index (GFI)	0.96	>0.90
Root Mean Square Error of Approximation (RMSEA)	0.06	<0.08

3.3. Effects of the Structural Model in SEM

Estimated unstandardized path coefficients depicted the effects among variables in the structural model (Figure 6). To facilitate comparison between relative magnitudes of effects, the width of each arrow in the diagram corresponds to the value of the respective standardized solution; that is, wider arrows indicate more significant standardized coefficients and thus stronger relative effects.

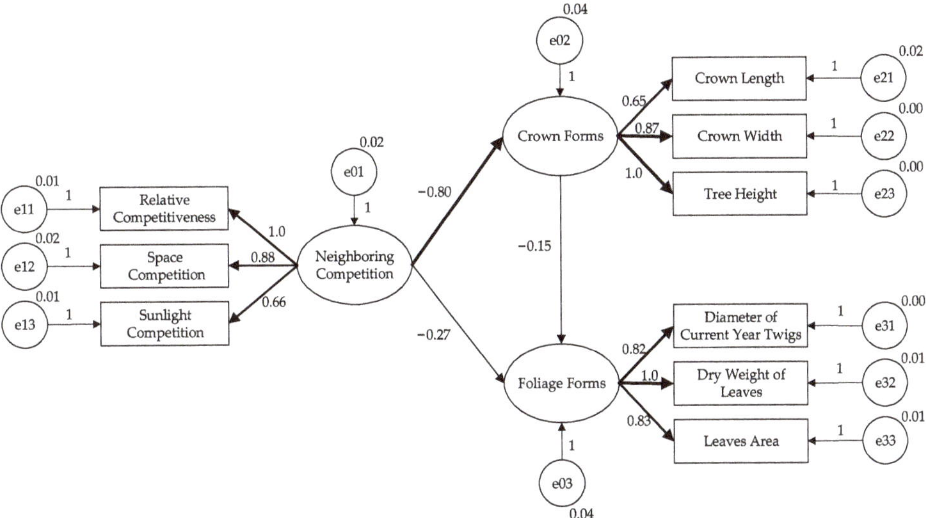

Figure 6. Structural equation model for the response of visual forms to the neighboring competition. Ovals and rectangles represent latent variables and observed variables, respectively. Circles represent the error variables. The numbers by the path are the unstandardized path factor loading. The arrows represent the cause-and-effect paths; the width of each arrow in the diagram corresponds to the value of the respective standardized solution.

Neighboring competition effects on crown forms were primarily direct, supported by a significant direct effect (-0.8, p-value < 0.001, Table 6). It means a 1% increase in neighboring competition directly reduced crown forms by 0.8%.

Table 6. Estimated total effects in the SEM, along with p-values (see Figure 6).

Paths in SEM		Direct Effect	Indirect Effect	Total Effect [1]	p-Value [2]
Neighboring Competition →	Crown Forms	−0.80	-- [3]	−0.80	0.000
	Foliage Forms	−0.27	0.12	−0.15	-- [3]
Crown Forms →	Foliage Forms	−0.15	-- [3]	−0.15	0.006

[1] Total effect was calculated as the sum of direct effect and indirect effect. [2] The p-value of direct path was calculated in *AMOS* version 20. The p-value of indirect path was not calculated. The p-value < 0.05 indicates a significant path. [3] The -- indicates no direct effect, no indirect effect, or no p-value for paths without a direct effect.

Neighboring competition affected foliage forms directly and indirectly, but the indirect effect mediated by crown forms was weaker than the direct effect. The indirect effect was estimated as the product of the direct effects of neighboring competition on crown forms and crown forms on foliage forms or approximately 0.12 (Figure 6, Table 6). In other words, a 1% increase in neighboring competition directly reduced foliage forms by 0.27% (Table 6) and indirectly increased by 0.12% ($0.12 = -0.27 \times -0.15$, Figure 6). Hence, the total effect of neighboring competition, which included the direct effect and all possible indirect effects, was negative (-0.15, p-value < 0.01, Table 6).

4. Discussion

Our findings demonstrated the potential relevance between visual forms of an individual tree and competitive pressure from its neighboring trees by means of the bivariate analysis and the SEM. *Cotinus coggygria* is an adaptive species that can modify its tree forms to improve sunlight capture and space occupation.

4.1. Effects of Neighboring Competition on Visual Forms

Our analysis of bivariate correlations between each primarily selected trait and neighboring competition indicators and the SEM analysis with three latent variables show that neighboring competition is the dominating driver of visual forms of individual trees for the analyzed data.

Crown forms were negatively affected by neighboring competition, supporting *Hypothesis 1* (Neighboring competition directly affects crown forms). We found an 80% decrease in crown forms with neighboring competitive pressure increase within the SEM. What is more, Pearson correlation coefficient for crown width with space competition ($r = -0.36$), crown length with relative competitiveness ($r = -0.56$), and tree height with sunlight competition ($r = -0.42$) indicated a solid negative relevance. Other research of tree growth and competition processes revealed similar negative correlations. González et al. [48] and Kahriman et al. [49] found that tree growth slows or halts when shaded by neighboring trees. Trees growing in completely open conditions with full lighting all around have crowns that may grow reflective of the trees' genetic disposition. However, Kramer et al. [50] found crown shape was affected when lower branches of shaded trees tend to become suppressed, decline, and die in the shade of upper branches of neighboring trees [15].

For foliage forms, neighboring competition has less significant total effects than crown forms, partially supporting *Hypothesis 2* (Neighboring competition affects foliage forms directly and indirectly mediated by crown forms). In the 218 competitive units, no significant trends were found between foliage form traits and relative competitiveness (the best R^2 is 0.01). Additionally, we got similar results for sunlight competition with the diameter of current year twigs and dry weight of leaves. All three competition indicators could not explain foliage form traits (the best R^2 is 0.03). Poorter et al. [51] found that environmental conditions have a significant impact on leaf traits. Leaves are affected not only by space [52] and light [53] but also by non-competition factors such as the soil nutrients (e.g., nitrogen, phosphorus, potassium), soil microbes, and organic matter [54,55]. However, these non-competition factors were outside the scope of this paper and not included in constructing the neighboring competition latent variable. Thus, their effects on foliage forms were not in consideration. Nevertheless, the error terms could partly account for these non-competition effects within the study site in SEM analysis. The direct effect of neighboring competition on foliage forms was -0.27, which indicates that foliage forms weakly negatively correlated with the neighboring competition. At the same time, the indirect effects mediated by crown forms were positive (0.12). It conformed to trade-offs between twigs and leaves, which interact with each other to improve light utilization and enhance defense capabilities [56]. In other words, leaves and twigs in large crowns shared crowded space and intercepted light. Thus, foliage forms negatively related to crown size. In SEM, the indirect effect is estimated as the product of the chain of direct effects [25].

Therefore, the positive indirect effects on foliage forms (0.12) were estimated as the product of negative effects of neighboring competition on crown forms (-0.8) and negative effects of crown forms on foliage forms (-0.15).

Moreover, the results completely supported *Hypothesis 3* (Crown forms and foliage forms respond differently to neighboring competition). Total effects of neighboring competition on crown forms and foliage forms varied by 81.3% (81.3% = $(0.8-0.15)/0.8 \times 100\%$). Thus, the difference would represent the distinct responses of different parts that affect visual forms to neighboring competition, even with measurement errors.

4.2. Application of SEM

Visual forms of individual trees were defined as a system in which crown forms and foliage forms interrelated and interacted with each other. Trees with the same crown forms may look different with different foliage forms (Figure 1). The bivariate analysis of each visual form trait and competition indicator could not reveal the complex inter-relationships. The best R^2 of bivariate correlations between foliage form traits and competition indicators are all minor and less than 0.05. In contrast, the direct effect of neighboring competition on foliage forms in the SEM indicated a slightly negative correlation. Furthermore, the indirect effect mediated by crown forms was eliminated in the bivariate analysis.

The realistic interrelationships between neighboring competition and visual forms could be complex and challenging to manipulate and test experimentally. The application of SEM made a way to understand the implications of neighboring competition on visual forms. Development of the "competition-visual form" model drew on published univariate models, theory, and expert knowledge to establish a set of working hypotheses on the interactions of different latent variables. Then, the hypotheses were translated into a mathematical model and fitted the latter to data collected from peri-urban forests dominated by *Cotinus coggygria*. The blend of confirmatory and exploratory SEM analyses provided insights into the complexity imposed by cascading effects of the different parts of trees.

The SEM was fitted by maximum likelihood. Because maximum likelihood was a complete information method, all processes and interactions represented in the model were considered simultaneously during model estimation [25]. This aspect of SEM was critical as the mechanisms for visual form responding to the competition were interactive and rarely dominated by one specific ecological factor [57]. As a result, sunlight attenuation and growing space competition were identified simultaneously as essential drivers of the observed patterns.

Unlike multiple regression, however, model parsimony was not the ultimate goal of SEM. Empirical re-specification of a model by dropping paths that were not significantly increased the chance of over-fitting a model to a specific data set. For example, the statistically insignificant path might be due to random variation or unique attributes of the analyzed dataset [25]. As mentioned above, a path was evaluated simultaneously in other paths; thus, its significance was meaningful only in the presence of other paths represented in the model. Therefore, empirical re-specification was not recommended, and a researcher should not feel compelled to drop insignificant paths from a model until replication of the results with other datasets or sufficient accumulation of evidence from related research indicated otherwise [25].

The SEM developed in this study explained competition effects and responses in terms of mechanisms that were not measured directly. For example, space competition was represented by simplified crown circle overlap area rather than a direct measurement of realistic crown dimensions of trees in competitive units. To the extent that more direct mechanisms and ultimate variables are becoming more directly observable, the dominating processes and mechanisms can be inferred with greater confidence.

4.3. Suggestions on Peri-Urban Forest Management

Forest landscapes are considered as a complex web of interactions whose threads are, for example, air, water, soil, vegetation, wildlife, insects, and microorganism-based.

Thus, it becomes challenging to focus on every specific individual. The management implications of visual forms and competition are profound. The SEM model proposed in this study explored the causal relationship between neighboring competition and visual forms, improving computer-based technology suitable for handling problems about forest visual resources, promoting qualitative management decision-making processes.

In peri-urban forests, neighboring competition is inevitable, but it can be managed. Thus, visual forms of trees can be adjusted by regulating competition relations by spacing through density management. As trees growing in open conditions, their dense crowns tend to be more visually attractive [58]. Reducing planting density by removing some diseased trees is beneficial to mature trees for crown form development. Thinning dominant trees with little aesthetic value in peri-urban forests could effectively allow sunlight to reach shaded trees and improve visual quality.

Since monotonous, repetitive visual forms are prone to aesthetic fatigue [59,60], increasing the diversity and difference of visual forms is an effective way to improve the visual attraction of peri-urban forests [61]. Adjusting competition intensity by punning branches and leaves will reshape crown forms and foliage forms. Promoting the health of trees by increasing the growing space and number of leaves would likely increase the visual quality of peri-urban forests [62].

The color of leaves is one of the essential characteristics of foliage forms and should be considered in the visual quality promotion. Leaves of *Cotinus coggygria* turn from green to red gradually when fall approaches, making a unique scenery of red leaves. Leaf color changing occurs mainly due to anthocyanin, temperature, and day length and has little to do with neighboring competition. However, increasing the area and number of leaves is still a reasonable choice to improve the visual quality of a peri-urban forest dominated by *Cotinus coggygria*. The data was collected in summer when appealing red leaves did not arise, and the color of leaves was not considered in the "competition-visual form" model. Still, the effects of color on visual appeal should be considered and studied [63] (pp. 149–186).

5. Conclusions

An exploratory attempt was made to apply SEM to understand visual forms of individual trees in a peri-urban forest landscape responding to the neighboring competition. As a multifaceted concept, the neighboring competition was constructed by space competition, sunlight competition, and relative competitiveness in this study. Moreover, visual forms of trees were analyzed from two different perspectives—the crown and foliage forms. The results indicated that neighboring competition, such as sunlight interception, space extrusion, dramatically affects the crown forms of trees. However, effects on foliage forms were relatively weaker than that on crown forms. Additionally, foliage forms and crown forms interact with each other. The SEM model and associated findings should facilitate peri-urban forest landscape management by providing insight into the causal mechanisms of visual forms of trees, thereby assisting in the visual quality promotion.

It is necessary to note that neighboring competition is not the only cause of variation in visual forms of trees in peri-urban forests. Factors such as soil, temperature, and moisture, would be worth a further investigation to quantify causal effects on visual forms of trees. However, these factors were outside the scope of this paper.

Author Contributions: Conceptualization, Y.C. and C.X.; methodology, Y.C. and C.X.; software, Y.C.; validation, Y.C. and C.X.; formal analysis, Y.C. and C.X.; investigation, Y.C. and J.Z.; resources, C.X.; data curation, Y.C.; writing—original draft preparation, Y.C.; writing—review and editing, Y.C. and R.J.H.; supervision, C.X.; project administration, Y.C., J.Z. and C.X. All authors have read and agreed to the published version of the manuscript.

Funding: This research was funded by the special programs for public welfare of Forestry in China, grant number 20140430102.

Data Availability Statement: The data presented in this study are available on request from the corresponding author. The data are not publicly available due to confidentiality.

Acknowledgments: Special thanks to the Short-term International Student Program for Postgraduates of Forestry First-Class Discipline (2019XKJS0501).

References

1. Haase, D.; Larondelle, N.; Andersson, E.; Artmann, M.; Borgström, S.; Breuste, J.; Gomez-Baggethun, E.; Gren, Å.; Hamstead, Z.; Hansen, R.; et al. A quantitative review of urban ecosystem service assessments: Concepts, models, and implementation. *Ambio* **2014**, *43*, 413–433. [CrossRef] [PubMed]
2. Schirpke, U.; Meisch, C.; Marsoner, T.; Tappeiner, U. Revealing spatial and temporal patterns of outdoor recreation in the European Alps and their surroundings. *Ecosyst. Serv.* **2018**, *31*, 336–350. [CrossRef]
3. Hartig, T.; Evans, G.W.; Jamner, L.D.; Davis, D.S.; Gärling, T. Tracking restoration in natural and urban field settings. *J. Environ. Psychol.* **2003**, *23*, 109–123. [CrossRef]
4. Deng, Z.J.; Cheng, H.Y.; Song, S.Q. Effects of temperature, scarification, dry storage, stratification, phytohormone and light on dormancy-breaking and germination of *Cotinus coggygria* var. *cinerea* (Anacardiaceae) seeds. *Seed Sci. Technol.* **2010**, *38*, 572–584. [CrossRef]
5. Zhang, Q. Basic Design the Dynamics of Visual Form from the First Generation Modernism Constructs the Master Reading the Building Constitute Beautiful. Master's Thesis, Tianjin University, Tianjin, China, 2006.
6. Dong, W.X. *Visual Form Semantics*; Shanghai University Press: Shanghai, China, 2007.
7. Li, W. Study on Skin Renewal Design of the Old Industrial Buildings. Master's Thesis, Qingdao University of Technological, Qingdao, China, 2012.
8. Rowe, N.; Speck, T. Plant growth forms: An ecological and evolutionary perspective. *New Phytol.* **2005**, *166*, 61–72. [CrossRef] [PubMed]
9. Umeki, K. A comparison of crown asymmetry between *Picea abies* and *Betula maximowicziana*. *Can. J. For. Res.* **1995**, *25*, 1876–1880. [CrossRef]
10. Getzin, S.; Wiegand, K. Asymmetric tree growth at the stand level: Random crown patterns and the response to slope. *For. Ecol. Manag.* **2007**, *242*, 165–174. [CrossRef]
11. Grams, T.E.E.; Andersen, C.P. Competition for resources in trees: Physiological versus morphological plasticity. *Prog. Bot.* **2007**, *68*, 356–381. [CrossRef]
12. Wu, J.; Chen, Y.; Liu, H.X.; Xu, L.J.; Jin, G.X.; Xu, C.Y. Effects of stand density and mingling intensity on tree morphology in natural scenic forest in Changbai Mountain. *Sci. Silvae Sin.* **2018**, *54*, 12–21. [CrossRef]
13. Kang, H.J.; Chen, Z.L.; Liu, P.; Zhang, Z.X.; Zhou, J.H. Intra-specific competition of *Emmenopterys henryi* and its accompanying species in the Dapanshan national nature reserve of Zhejiang province. *Acta Ecol. Sin.* **2008**, *28*, 3456–3463. [CrossRef]
14. Oheimb, G.V.; Lang, A.C.; Bruelheide, H.; Forrester, D.I.; Wäsche, I.; Yu, M.J.; Härdtle, W. Individual-tree radial growth in a subtropical broad-leaved forest: The role of local neighbourhood competition. *For. Ecol. Manag.* **2011**, *261*, 499–507. [CrossRef]
15. Yin, D.S.; Ge, W.Z.; Zhang, F.H.; Shen, H.L. Competition relationship of populations of natural secondary *Acer mono* forest. *Bull. Bot. Res.* **2012**, *32*, 105–109.
16. Yu, L.; Song, M.Y.; Lei, Y.B.; Korpelainen, H.; Niinemets, Ü.; Li, C.Y. Effects of competition and phosphorus fertilization on leaf and root traits of late-successional conifers *Abies fabri* and *Picea brachytyla*. *Environ. Exp. Bot.* **2019**, *162*, 14–24. [CrossRef]
17. Grace, J.B.; Youngblood, A.; Scheiner, S.M. Structural equation modeling and ecological experiments. In *Real World Ecology: Large-Scale and Long-Term Case Studies and Methods*; Miao, S.L., Carstenn, S., Nungesser, M., Eds.; Springer: New York, NY, USA, 2009; pp. 19–46.
18. Shipley, B. *Cause and Correlation in Biology: A User's Guide to Path Analysis, Structural Equations and Causal Inference with R.*; Cambridge University Press: Cambridge, UK, 2016.
19. Grace, J.B. Structural equation modeling for observational studies. *J. Wildl. Manag.* **2008**, *72*, 14–22. [CrossRef]
20. Timilsina, N.; Escobedo, F.J.; Staudhammer, C.L.; Brandeis, T. Analyzing the causal factors of carbon stores in a subtropical urban forest. *Ecol. Complex.* **2014**, *20*, 23–32. [CrossRef]
21. Cottam, G.; Curtis, J.T. The use of distance measures in phytosociological sampling. *Ecology* **1956**, *37*, 451–460. [CrossRef]
22. Zhang, J.E. *Experimental Research Methods and Techniques Are Commonly Used in Ecology*; Chemical Industry Press: Beijing, China, 2007; pp. 86–87.
23. Schumacker, R.E.; Lomax, R.G. *A Beginner's Guide to Structural Equation Modeling*; Routledge: New York, NY, USA, 2010.
24. Lam, T.Y.; Maguire, D.A. Structural equation modeling: Theory and applications in forest management. *Int. J. Forestry Res.* **2012**, 1–16. [CrossRef]
25. Kline, R.B. *Principles and Practice of Structural Equation Modeling*, 3rd ed.; The Guilford Press: New York, NY, USA, 2011.
26. Roxo, J. Elements of Art: Interpreting Meaning through the Language of Visual Cues. Doctoral Thesis, State University of New York at Stony Brook, New York, NY, USA, 2018.

27. Iwasa, Y.; Cohen, D.; Leon, J.A. Tree height and crown shape, as results of competitive games. *J. Theor. Biol.* **1984**, *112*, 279–297. [CrossRef]
28. Lindh, B.C.; Gray, A.N.; Spies, T.A. Responses of herbs and shrubs to reduced root competition under canopies and in gaps: A trenching experiment in old-growth Douglas-fir forests. *Can. J. Forest Res.* **2003**, *33*, 2052–2057. [CrossRef]
29. Riegel, G.M.; Miller, R.F.; Krueger, W.C. Competition for resources between understory vegetation and overstory *Pinus ponderosa* in northeastern Oregon. *Ecol. Appl.* **1992**, *2*, 71–85. [CrossRef] [PubMed]
30. Olivier, M.D.; Robert, S.; Fournier, R.A. Response of sugar maple (*Acer saccharum*, Marsh.) tree crown structure to competition in pure versus mixed stands. *For. Ecol. Manag.* **2016**, *374*, 20–32. [CrossRef]
31. Lieffers, V.J.; Messier, C.; Stadt, K.J.; Gendron, F.; Comeau, P.G. Predicting and managing light in the understory of boreal forests. *Can. J. For. Res.* **1999**, *29*, 796–811. [CrossRef]
32. Bella, I.E. A new competition model for individual trees. *For. Sci.* **1971**, *17*, 364–372.
33. Opie, J.E. Predictability of individual tree growth using various definitions of competing basal area. *For. Sci.* **1968**, *14*, 314–323. [CrossRef]
34. Smith, W.R.; Farrar, R.M., Jr.; Murphy, P.A.; Yeiser, J.L.; Meldahl, R.S.; Kush, J.S. Crown and basal area relationships of open-grown southern pines for modeling competition and growth. *Can. J. For. Res.* **1992**, *22*, 341–347. [CrossRef]
35. Van der Zee, J.; Lau, A.; Shenkin, A. Understanding crown shyness from a 3-D perspective. *Ann. Bot.* **2021**, *3*, 1–11. [CrossRef]
36. Onoda, Y.; Bando, R. Wider crown shyness between broad-leaved tree species than between coniferous tree species in a mixed forest of *Castanopsis cuspidata* and *Chamaecyparis obtusa*. *Ecol. Res.* **2021**, *36*, 733–743. [CrossRef]
37. Smith, S.H.; Bell, J.F. Using competitive stress index to estimate diameter growth for thinned Douglas-fir stands. *For. Sci.* **1983**, *29*, 491–499. [CrossRef]
38. Baskerville, G.L. Dry-matter production in immature balsam fir stands: Roots lesser vegetation, and total stand. *For. Sci.* **1966**, *12*, 49–53. [CrossRef]
39. Weiner, J.; Damgaard, C. Size-asymmetric competition and size-asymmetric growth in a spatially explicit zone-of-influence model of plant competition. *Ecol. Res.* **2006**, *21*, 707–712. [CrossRef]
40. Horn, H.S. *The Adaptive Geometry of Trees*; Princeton University Press: Princeton, NJ, USA, 1971.
41. Westoby, M.; Wright, I.J. The leaf size-twig size spectrum and its relationship to other important spectra of variation among species. *Oecologia* **2003**, *135*, 621–628. [CrossRef]
42. Niklas, K.J. Size-dependent allometry of tree height, diameter and trunk-taper. *Ann. Bot.* **1995**, *75*, 217–227. [CrossRef]
43. Tyree, M.T.; Ewers, F.W. The hydraulic architecture of trees and other woody plants. *New Phytol.* **1991**, *119*, 345–360. [CrossRef]
44. Weiner, J.; Berntson, G.M.; Thomas, S.C. Competition and growth form in a woodland annual. *J. Ecol.* **1990**, *78*, 459–469. [CrossRef]
45. Dell'Olio, L. Structural equation models. In *Public Transportation Quality of Service*; Dell'Olio, L., Ibeas, A., de Oña, J., de Oña, R., Eds.; Elsevier: Amsterdam, The Netherlands, 2018; pp. 141–152. [CrossRef]
46. Browne, M.W.; Cudeck, R. Alternative ways of assessing model fit. *Sociol. Method. Res.* **1992**, *21*, 230–258. [CrossRef]
47. Gerbing, D.W.; Anderson, J.C. Monte Carlo evaluations of goodness of fit indices for structural equation models. *Sociol. Method. Res.* **1992**, *21*, 132–160. [CrossRef]
48. González de Andrés, E.; Camarero, J.J.; Blanco, J.A.; Imbert, J.B.; Lo, Y.H.; Sangüesa-Barreda, G.; Castillo, F.J. Tree-to-tree competition in mixed European beech-*Scots pine* forests has different impacts on growth and water-use efficiency depending on site conditions. *J. Ecol.* **2018**, *106*, 59–75. [CrossRef]
49. Kahriman, A.; Şahin, A.; Sönmez, T.; Yavuz, M. A novel approach to selecting a competition index: The effect of competition on individual-tree diameter growth of Calabrian pine. *Can. J. For. Res.* **2018**, *48*, 1217–1226. [CrossRef]
50. Kramer, R.D.; Sillett, S.C.; Van Pelt, R.; Franklin, J.F. Neighborhood competition mediates crown development of *Picea sitchensis* in Olympic rainforests: Implications for restoration management. *For. Ecol. Manag.* **2019**, *441*, 127–143. [CrossRef]
51. Poorter, H.; Niinemets, Ü.; Poorter, L.; Wright, I.; Villar, R. Causes and consequences of variation in leaf mass per area (LMA): A meta-analysis. *New Phytol.* **2009**, *182*, 565–588. [CrossRef]
52. Lee, T.D.; Bazzaz, F.A. Effects of defoliation and competition on growth and reproduction in the annual plant *Abutilon theophrasti*. *J. Ecol.* **1980**, *68*, 813–821. [CrossRef]
53. Mekonnen, Z.A.; Riley, W.J.; Grant, R.F. Accelerated nutrient cycling and increased light competition will lead to 21st century shrub expansion in North American Arctic tundra. *J. Geophys. Res. Biogeo.* **2018**, *123*, 1683–1701. [CrossRef]
54. Longstreth, D.J.; Nobel, P.S. Nutrient influences on leaf photosynthesis: Effects of nitrogen, phosphorus, and potassium for *Gossypium hirsutum* L. *Plant Physiol.* **1980**, *65*, 541–543. [CrossRef]
55. Reich, P.B.; Ellsworth, D.S.; Walters, M.B. Leaf structure (specific leaf area) modulates photosynthesis-nitrogen relations: Evidence from within and across species and functional groups. *Funct. Ecol.* **1998**, *12*, 948–958. [CrossRef]
56. Long, J.Y.; Zhao, Y.M.; Kong, X.Q.; Chen, Z.Y.; Wang, X.S.; Zhao, K.; Cao, R.; Huang, L.S.; Lü, J.; Cui, Y.; et al. Trade-offs between twig and leaf traits of ornamental shrubs grown in shade. *Acta Ecol. Sin.* **2018**, *38*, 1–9. [CrossRef]
57. Kocher, S.D.; Harris, R. *Forest Stewardship Series 5: Tree Growth and Competition*; University of California Agriculture and Natural Resources: California, CA, USA, 2007; pp. 1–10.
58. Coomes, D.A.; Allen, R.B. Testing the metabolic scaling theory of tree growth. *J. Ecol.* **2009**, *97*, 1369–1373. [CrossRef]
59. Lai, Y.B. On the phenomenon of "aesthetic fatigue" in the perspective of consumer society. *Mov. Lit.* **2011**, *21*, 17–18.
60. Feng, X.L. *Aesthetics in the System of Human Life*; Anhui Education Press: Hefei, China, 2004.

61. Xue, X. A Study on Aesthetic Fatigue of Forest Landscapes of Tourists Based on EEG. Master's Thesis, Central South University of Forestry and Technology, Changsha, China, 2015.
62. Hull, R.B.; Buhyoff, G.J. The scenic beauty temporal distribution method: An attempt to make scenic beauty assessments compatible with forest planning efforts. *For. Sci.* **1986**, *32*, 271–286.
63. Bell, S.; Blom, D.; Rautamäki, M.; Castel-Branco, C.; Simson, A.; Olsen, I.A. Design of urban forests. In *Urban Forests and Trees: A Reference Book*; Konijnendijk, C.C., Nilsson, K., Randrup, T.B., Schipperijn, J., Eds.; Springer: Berlin/Heidelberg, Germany, 2005; pp. 149–186.

Article

The Temporal Variation of the Microclimate and Human Thermal Comfort in Urban Wetland Parks: A Case Study of Xixi National Wetland Park, China

Zhiyong Zhang [1], Jianhua Dong [2], Qijiang He [2] and Bing Ye [1,*]

[1] Research Institute of Forestry Policy and Information, Chinese Academy of Forestry, Beijing 100091, China; zhangzhiyong@caf.ac.cn

[2] Hangzhou Academy of Forestry, Hangzhou 310022, China; jianhuadong@126.com (J.D.); yongkobe666@163.com (Q.H.)

* Correspondence: yb70@caf.ac.cn

Abstract: As an important part of the ecological infrastructure in urban areas, urban wetland parks have the significant ecological function of relieving the discomfort of people during their outdoor activities. In recent years, the specific structures and ecosystem services of urban wetland parks have been investigated from different perspectives. However, the microclimate and human thermal comfort (HTC) of urban wetland parks have rarely been discussed. In particular, the changing trends of HTC in different seasons and times have not been effectively presented. Accordingly, in this research, a monitoring platform was established in Xixi National Wetland Park, China, to continually monitor its microclimate in the long term. Via a comparison with a control site in the downtown area of Hangzhou, China, the temporal variations of the microclimate and HTC in the urban wetland park are quantified, and suggestions for clothing are also provided. The results of this study demonstrate that urban wetland parks can mitigate the heat island effect and dry island effect in summer. In addition, urban wetland parks can provide ecological services at midday during winter to mitigate the cold island effect. More importantly, urban wetland parks are found to exhibit their best performance in improving HTC during the daytime of the hot season and the midday period of the cold season. Finally, the findings of this study suggest that citizens should take protective measures and enjoy their activities in the morning, evening, or at night, not at midday in hot weather. Moreover, extra layers are suggested to be worn before going to urban wetland parks at night in cold weather, and recreational activities involving accommodation are not recommended. These findings provide not only basic scientific data for the assessment of the management and ecological health value of Xixi National Wetland Park and other urban wetland parks with subtropical monsoon climates, but also a reference for visitor timing and clothing suggestions for recreational activities.

Keywords: microclimate; human thermal comfort; outdoor thermal environment; public health; ecological services

Citation: Zhang, Z.; Dong, J.; He, Q.; Ye, B. The Temporal Variation of the Microclimate and Human Thermal Comfort in Urban Wetland Parks: A Case Study of Xixi National Wetland Park, China. *Forests* **2021**, *12*, 1322. https://doi.org/10.3390/f12101322

Academic Editors: Manuel Esperon-Rodriguez and Tina Harrison

Received: 27 July 2021
Accepted: 24 September 2021
Published: 28 September 2021

Publisher's Note: MDPI stays neutral with regard to jurisdictional claims in published maps and institutional affiliations.

1. Introduction

Due to the complex social process of rapid urbanization, approximately half of the global population lives in urban areas, and this percentage is still increasing [1,2]. Undeniably, modernized fundamental infrastructure has provided great convenience to human life. Compared to a century ago, people living in urban areas work efficiently and create high economic value. However, due to rapid urbanization, factors such as the expansion of urban areas, the use of concrete and asphalt, and the loss of natural resources and space have led to increased and decreased temperatures in urban areas in summer and winter, respectively, as well as reduced humidity, compared to rural areas. These phenomena are respectively known as the urban heat (cold) island effect and the urban dry island effect [2]. Furthermore, in the context of global climate change, extreme weather events, including urban waterlogging, dry winds, and persistent heat and cold waves, are frequently observed in urban areas [3,4]. This can be attributed to the spatial heterogeneity and fragmentation of the urban environment. Human-induced urban spaces cause damage to the essential functions of natural ecosystems, including substance circulation, energy exchange, and information transfer. Indeed, urban ecosystems cannot regulate their temperature independently, as the concrete ground is directly exposed to sunlight, and air conditioners and other electrical devices are widely used [5]. This results in the incapability of the urban ecosystem to effectively exert its ecological function of adjusting the microclimate, leading to the growing discomfort of people during their outdoor activities [6].

The ecological infrastructure plays an important role in maintaining the integrity of the urban ecological structure. Due to the COVID-19 pandemic, people are required to maintain social distance. In this case, there has been more awareness of the significance of natural outdoor space and fresh air. As one of the most important ecological infrastructures in urban areas, wetland parks provide various ecosystem services, including climate regulation, flood control, aquifer recharging, water purification, carbon sequestration, and they act as habitats for plants and animals [7–9]. In recent years, the specific structures and ecosystem services of wetland parks have been investigated from different perspectives [10–12]. The human thermal comfort (HTC) of the environment is also an important component of ecosystem services and plays a vital role in improving the physical and mental health of visitors in wetland parks. Unfortunately, the HTC of urban wetland parks has rarely been discussed.

HTC, which is a concept that reflects humans' perception of the thermal environment, has been widely employed to evaluate the comfort level of the body in response to the environment [13,14]. The traditional method of the measurement of HTC is to use a series of meteorological factors (e.g., temperature, humidity, wind speed, radiation, etc.) to establish a comprehensive index by which to quantify the bioclimatic conditions for human. Over the past century, more than 100 indices have been developed and used to assess HTC in combination with meteorological factors [15]. In recent years, several models, such as the urban canopy model, Urban Tethys-Chloris (UT&C), have been proposed to evaluate the urban thermal climate. UT&C is a combination of an urban canyon scheme and an ecohydrological model, and considers air and surface temperatures, air humidity, and soil moisture, as well as the urban energy and hydrological fluxes in the absence of snow [16–18]. These indexes and models all provide the basis for assessing the human thermal sensation of the environment. Previous studies of microclimates and HTC focused on urban outdoor spaces, college campuses, street green spaces, and city parks [19–23]. However, most of these studies were conducted in the summer, and the monitoring periods tended to be short (typically 3–7 d). Additionally, some studies did not consider the effects of relative humidity and wind speed on HTC, and continual monitoring in the long term was not sufficient, leading to the ineffective presentation of the changing trends of HTC in different seasons and at different times.

Therefore, based on the annual monitoring of the microclimate features and HTC of Xixi National Wetland Park, China in different seasons, this study determines the period of time in a day when people can obtain the highest level of HTC during recreational activities, and offers suggestions for appropriate clothing. The results provide a reference for the design and management of ecology therapy activities in wetland parks.

2. Target Site and Methods

2.1. Target Site

In this study, Xixi National Wetland Park in China was selected as the target site. Xixi National Wetland Park (30°15.39′ N–30°16.96′ N, 120°03.16′ E–120°04.94′ E) is 16 km from the downtown area of Hangzhou, China. As a typical urban wetland park and China's first national wetland park, Xixi National Wetland Park was established by the State Forestry and Grassland Administration (formerly the State Forestry Administration) of China in 2005 [24]. It has a total area of 11.5 km^2, and over 50% of its surface is covered by water. The park has a typical humid subtropical monsoon climate with high temperatures and precipitation in summer, and low precipitation in winter. The average temperature is about 16.2 °C, and the average precipitation is about 1400 mm per year. Xixi National Wetland Park consists of three areas: the wetland ecological protection and cultivation area (east), the wetland eco-tourism leisure area (center), and the wetland ecological landscape conservation area (west). As shown in Figure 1, the wetland park site (WPS) and the control site (CS) are in the ecological protection and cultivation area and the downtown area of Hangzhou, respectively. When the positioning monitoring platform used in this study was installed, the destruction of vegetation around the site was minimized. After installation, the original vegetation was reconstructed. Excluding the presence of protective measures, such as a guardrail installed around the site, there is no man-made shade or open space to prevent the modification of wind flow and the limiting of heat dissipation. The CS is not shaded by nearby buildings, and is not affected by wind channelling, downdraft effects, etc. Thus, the microclimate characteristics of the regions represented by the two platforms are objectively reflected.

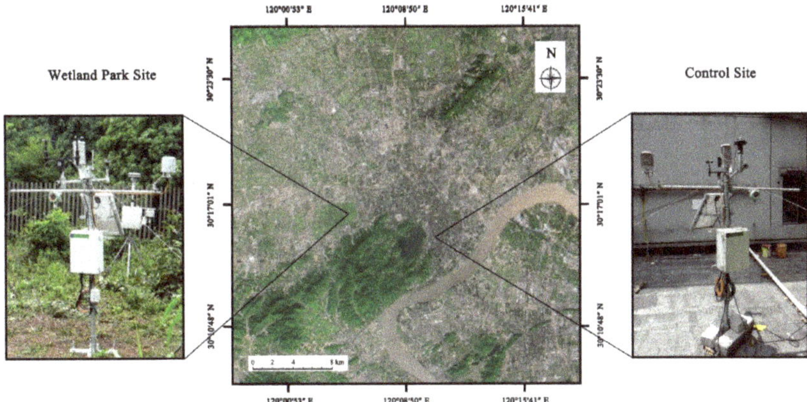

Figure 1. The wetland park site and the control site in Hangzhou, China.

2.2. Methods

2.2.1. Monitoring and Data Sources

At each site, a monitoring platform was set up to monitor the air temperature, relative humidity, and wind speed, as shown in Figure 1. The air temperature and relative humidity were detected by a sensor (Vaisala HMP155, Vaisala, Inc., Woburn, USA), and the wind speed was measured by an anemometer (RM Young Wind Sentry Set03002-1, Campbell Scientific Inc., Logan, USA) 24 h per day.

The data collector (CR1000, Campbell Scientific Inc., Logan, USA) recorded data every 15 min and transmitted it to a remotely controlled computer via the MA8-L GPRS terminal. The description of measuring devices is provided in Table 1. In this study, data collected from January–December 2016 were used.

Table 1. The description of measuring sensors.

Sensors	Product Type	Work Environment	Measuring Range	Accuracy
Temperature and humidity sensor	HMP155		−40–60 °C, 0–60 %	±0.2 °C, ±1%
Anemometer	Set03002-1	−20–50 °C	0–60 m/s	±0.5 m/s
Data collector	CR1000	−20–65 °C		

2.2.2. Assessment Indicators of HTC

Table 1 reports the common indices of HTC used internationally, each of which has its own applicable weather conditions. By comparing various indices, Blazejczyk et al. confirmed that each index has a high correlation with each other (the lowest R^2 coefficient was 93.1), regardless of the inclusion of a radiation factor in the equation. Moreover, it was pointed out that non-radiation equations are used more widely in some climatic conditions [15]. In other words, each index demonstrates high consistency in the assessment of HTC. Therefore, in the application of bioclimatic indices for the assessment of HTC, the climatic conditions of the research area and people's thermal perception habits should be fully considered during index selection.

Similar to the indices listed in Table 2, the indices reported in Table 3 are the most widely used in China. In comprehensive consideration of the scope of application of each index and the seasonal dressing habits of the Chinese people, the human thermal comfort index (HTCI) and clothing thickness index (CTI) were employed to evaluate the human perception of the environment. The HTCI was focused on evaluating the comfort level of the body. Additionally, the CTI can serve as an indicator of clothing.

Table 2. The combined environmental variables and applicable weather conditions of indices.

Indices	Combined Environmental Variables	Applicable Weather Conditions	References
Heat index (HI)	Air temperature and relative humidity	Hot weather	Steadman [25]
Humidex (HD)	Air temperature and air vapor pressure	Hot weather	Sirangelo et al. [26] and Masterton and Richardson [27]
Effective temperature (ET)	Air temperature, relative humidity, and wind speed	Hot and warm weather	Houghton and Yaglou [28]
Wet-bulb globe temperature (WBGT)	Air temperature, globe temperature, and natural wet bulb temperature	Hot weather	ISO 7243 [29] and d'Ambrosio Alfano et al. [30]
Wind chill index (WCI)	Air temperature and wind speed	Cold weather	Lin et al. [31] and Blazejczyk et al. [15]
Standard effective temperature (SET)	Air temperature, relative humidity, and mean radiant temperature	Hot weather	Gagge et al. [32] and Gagge et al. [33]
Insulation required (IREQ)	Air temperature, mean radiant temperature, relative humidity, and air velocity	Cold weather	ISO 11079 [34] and d'Ambrosio Alfano et al. [35]
Predicted heat strain (PHS)	Air temperature, mean radiant temperature, relative humidity, and air velocity	Hot weather	ISO 7933 [36] and d'Ambrosio Alfano et al. [37]
Physiological equivalent temperature (PET)	Air temperature, air vapor pressure, wind speed, and mean radiant temperature	Hot and cold weather	Höppe [38]
Universal Thermal Climate Index (UTCI)	Air temperature, mean radiant temperature, air vapor pressure, and wind speed	Hot and cold weather	Jendritzky et al. [39]

Table 3. The most widely used indices in China.

Indices	Combined Environmental Variables	Equations *	Applicable Weather Conditions
Human thermal comfort index (HTCI)	Air temperature, relative humidity, and wind speed	$\text{HTCI} = 0.6 \times (|T - 24|) + 0.07 \times (|RH - 70|) + 0.5 \times (|V - 2|)$	Hot and cold weather
Thermal humidity index (THI)	Air temperature and relative humidity	$\text{THI} = T - 0.55 \times (1 - RH)(T - 14.5)$	Hot weather

Table 3. *Cont.*

Indices	Combined Environmental Variables	Equations*	Applicable Weather Conditions
Human comfort index (HCI)	Air temperature, relative humidity, and wind speed	$HCI = 1.8T - 0.55 \times (1.8T - 26)(1 - RH) - 3.2 \times \sqrt{V} + 32$	Hot and cold weather
Discomfort index (DI)	Air temperature and relative humidity	$DI = 1.8T + 32 - 0.55 \times (1 - RH)(1.8T - 26)$	Hot weather
Cool index (CI)	Air temperature, and wind speed	$CI = \left(10 \times \sqrt{V} + 10.45 - V\right)(33 - T)$	Cold weather

* T is the temperature (°C), RH is the relative humidity (%), and V is the wind speed (m/s).

(1) The HTCI, proposed by Lu et al. [40], is a synthesis of temperature, relative humidity, and wind speed, and is calculated by the following Equation (1) [41,42]. Compared to the thermal humidity index (which is only applicable to hot climate conditions), the HTCI can better reflect the outdoor satisfaction level of humans in the four seasons, and thus it is widely applied.

$$HTCI = 0.6 \times (|T - 24|) + 0.07 \times (|RH - 70|) + 0.5 \times (|V - 2|), \tag{1}$$

where T is the temperature (°C), RH is the relative humidity (%), and V is the wind speed (m/s). The HTCI is negatively related to the body perception of comfort, as presented in Table 4.

Table 4. The relationship between the HTCI and human body perception [41,42].

Level	Range	Feeling
I	$HTCI \leq 4.55$	Very comfortable
II	$4.55 < HTCI \leq 6.95$	Comfortable
III	$6.95 < HTCI \leq 9.00$	Uncomfortable
IV	$HTCI > 9.00$	Very uncomfortable

(2) The CTI is also a synthesis of temperature, relative humidity, and wind speed, and is calculated by the following equations [41,43].
If $T \leq 18\,°C$ and $RH \geq 60\%$,

$$CTI = \frac{\left[1 + \frac{0.4(RH - 60)}{100}\right] \times 0.61(33 - T)}{(1 - 0.01165V^2)}. \tag{2}$$

If $18\,°C < T < 26\,°C$ or $T \leq 18\,°C$, and $RH < 60\%$, or, if $T \geq 26\,°C$ and $RH < 60\%$,

$$CTI = \frac{0.61(33 - T)}{(1 - 0.01165V^2)}. \tag{3}$$

If $T \geq 26\,°C$ and $RH \geq 60\%$,

$$CTI = \frac{\left[1 - \frac{0.4(RH - 60)}{100}\right] \times 0.61(33 - T)}{(1 - 0.01165V^2)}. \tag{4}$$

In these equations, T is the temperature (°C), RH is the relative humidity (%), and V is the wind speed (m/s). Table 5 presents the relationship between the CTI and the clothing suggestion, which was adapted from the work of Gu et al. [41] and Zhu et al. [43].

Table 5. Division of the CTI and clothing suggestion.

Level	Range	Clothing Suggestion
I	$CTI \leq 1.5$	Short-sleeve-based summer clothing
II	$1.5 < CTI \leq 6$	Long-sleeve shirt
III	$6 < CTI \leq 11$	Shirts plus a jacket or suit
IV	$11 < CTI \leq 18$	Sweater plus a thin coat or thin cotton coat
V	$CTI > 18$	Thick sweater plus a wool coat or down jacket or other winter clothing

2.2.3. Data Analysis

Herein, spring refers to the months of March, April, and May, summer includes June, July, and August, autumn includes September, October, and November, and winter includes December, January, and February. The winter data considered in this study were collected in December, January, and February 2016.

When the observation data were obtained from the data collector, a quality check was carried out first. After removing the abnormal data, statistical analysis was performed. The temperature, relative humidity, and wind speed at the same time each day of the year were averaged to obtain microclimatic data at that time, from which the daily variations are plotted. After that, the temperature, relative humidity and wind speed in each month and season were averaged arithmetically, i.e., the microclimatic data of the month and season were obtained respectively. Finally, the daily, monthly, and seasonal variations of HTCI and CTI were calculated according to Equations (1)–(4). The significance of arithmetic mean values for all data sets was assessed using Student's *t*-test. Error bars were calculated with the s.d. function. The differences at $p < 0.05$ were considered statistically significant by One-way ANOVA. According to the Student's *t*-test, characters in the figure represent statistically significant differences compared with control (* $p < 0.05$, ** $p < 0.01$ and *** $p < 0.001$). Data analysis was conducted using SPSS 19.0 software (IBM® SPSS® software, Armonk, NY, USA), and the data were plotted by GraphPad Prism 8 software (GraphPad Software, Inc., San Diego, CA, USA).

3. Results

3.1. Seasonal, Monthly, and Daily Variations of the Microclimate

3.1.1. Seasonal, Monthly, and Daily Variations of Temperature

The seasonal variations of average temperature in both the WPS and CS exhibited a greater range in summer (27.5 and 29.7 °C) than in autumn (18.9 and 20.4 °C), spring (16.6 and 18.1 °C), and winter (7.0 and 8.5 °C) (Figure 2a). The monthly variation of the average temperature could be described as a unimodal curve, and in July and August, the peak values were reached in both the WPS and CS (Figure 2b). Regarding the daily variation, the average temperature exhibited a trend from decreasing to increasing, reached the highest point between 12:00 and 16:00, and then gradually decreased (Figure 2c,d and Supplementary Figure S1a–j). The seasonal, monthly, and daily variations of the temperature followed the natural laws of cool temperatures in the winter, warm temperatures in the summer, high temperatures during the day, and low temperatures at night. Moreover, the average temperature of the WPS in each season and month was significantly lower than that of the CS, and the percentage of reduction was between 6.30% and 20.83%. However, in terms of the daily variation, some interesting findings were discovered. Taking February as an example (Figure 2c), the average temperature of the WPS was higher than that of the CS between 11:45 and 16:30. This also occurred in March, January, and December (Supplementary Figure S1a,b,j). In other months, the average temperature of the WPS was lower than that of the CS in the 24-h cycle (Supplementary Figure S1c–i). Compared to the urban environment, the ecological infrastructures in the wetland park can reform the wind pattern and limit heat dissipation at noon in the cold season, so that more heat can be stored in the environment. Figure 2d presents the daily variation of temperature in July.

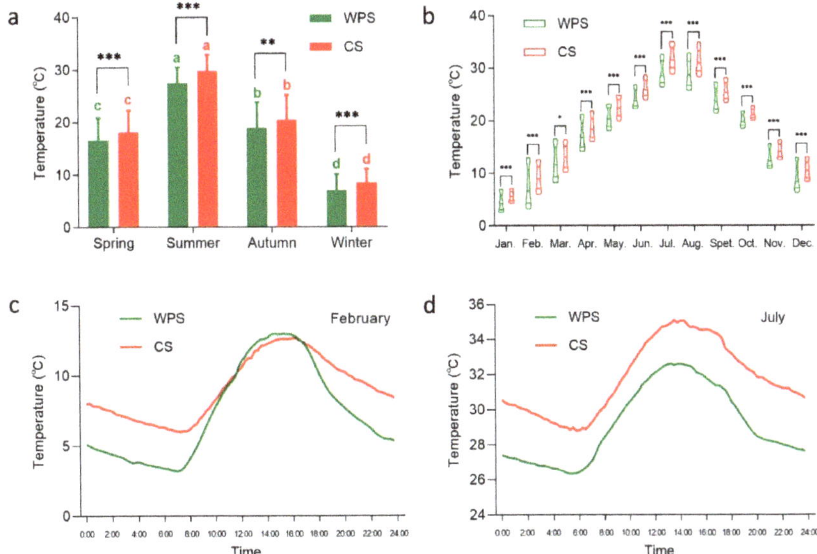

Figure 2. The seasonal, monthly, and daily variations of the average temperature in the WPS and CS. (**a**) The seasonal variation of the average temperature; different lowercase letters of the same color indicate results of a one-way ANOVA followed by Tukey's test ($p < 0.05$), and error bars represent the standard deviation. (**b**) The monthly variation of the average temperature. (**c**) The daily variation of temperature in February. (**d**) The daily variation of temperature in July. Note: Significant differences between the means of WPS and CS were determined using Student's *t*-test (* $p < 0.05$, ** $p < 0.01$, *** $p < 0.001$).

3.1.2. Seasonal, Monthly, and Daily Variations of Relative Humidity

The average relative humidity in both the WPS and CS differed with various seasons. It peaked in autumn (85.45% and 75.03%), followed by summer (83.34% and 68.74%) and spring (79.30% and 67.97%), and the lowest average relative humidity occurred in winter (77.08% and 65.32%). The average relative humidity of the WPS presented significant differences between the four seasons, while that of the CS exhibited no significant difference between spring and summer and had a significant difference with other seasons (Figure 3a). Regarding the monthly variation, the relative humidity was at a relatively high level in all months, there were negligible changes excluding February and August (Figure 3b). The daily variation of the average relative humidity first increased and then decreased, reached the lowest point at 14:40–16:00, and then increased gradually. In February, the average relative humidity of the WPS and CS was relatively close, between 11:45 and 16:30, while the percentage increase of relative humidity in the WPS was less than 10% that in the CS (Figure 3c). This was also observed in January, March, and December (Supplementary Figure S2a,b,j). In other months, the percentage increase of relative humidity was generally above 10% most of the day and overnight (Supplementary Figure S1c–i). Figure 3d presents the daily variation of the relative humidity in August. The average relative humidity of the WPS and CS exhibited significant differences on monthly and annual scales. Moreover, the average relative humidity of the WPS in each season and month was significantly higher than that of the CS, and the percentage increase was between 12.77% and 24.77%.

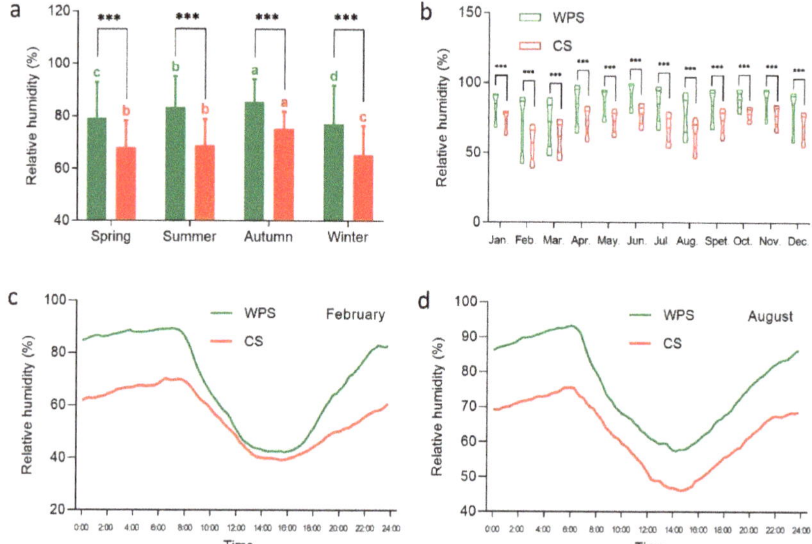

Figure 3. The seasonal, monthly, and daily variations of the average relative humidity in the WPS and CS. (**a**) The seasonal variation of the average relative humidity; different lowercase letters of the same color indicate results of a one-way ANOVA followed by Tukey's test ($p < 0.05$), and error bars represent the standard deviation. (**b**) The monthly variation of the average relative humidity. (**c**) The daily variation of relative humidity in February. (**d**) The daily variation of relative humidity in August. Note: Significant differences between the means of WPS and CS were determined using Student's *t*-test (*** $p < 0.001$).

3.1.3. Seasonal, Monthly, and Daily Variations of Wind Speed

The wind speed in both the WPS and CS exhibited seasonal variation, and the wind speed in the WPS was significantly lower than that in the CS. The highest wind speed in the WPS occurred in winter (0.11 m/s), followed by summer (0.08 m/s) and spring (0.07 m/s), and the lowest wind speed occurred in autumn (0.02 m/s). The wind speeds in spring and summer were not substantially different, but recorded a large difference compared with winter and autumn ($p < 0.05$). The highest wind speed in the CS occurred in summer (0.21 m/s), followed by spring (0.14 m/s), and the lowest wind speed occurred in winter and autumn (0.12 m/s), which demonstrated a significant variation as compared with spring and summer ($p < 0.05$) (Figure 4a).

The monthly wind speed in the WPS and CS displayed fluctuations. In general, the wind speed in the WPS was relatively high in December, January, February, March, and June, while that in the CS was relatively high in February, March, June, July, and August, and the dispersion of the wind speed value in the CS was relatively large (Figure 4b). The daily variation curve for each month reveals that the wind speed in both the WPS and CS was the highest between 14:00 and 16:00 with a single peak. The daily variation of the wind speed in the CS was large in each month, whereas that in the WPS exhibited small fluctuations between April and November (Supplementary Figure S3c–i) and large fluctuations in other months (Supplementary Figure S3a,b,j). Figure 4c,d presents the daily variation of wind speed in February and August.

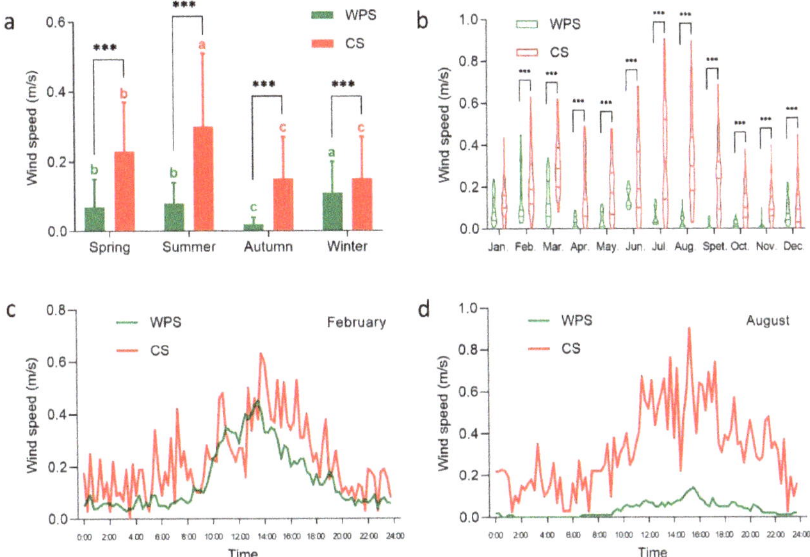

Figure 4. The seasonal, monthly, and daily variations of the average wind speed in the WPS and CS. (a) The seasonal variation of the average wind speed; different lowercase letters of the same color indicate results of a one-way ANOVA followed by Tukey's test ($p < 0.05$), and error bars represent the standard deviation. (b) The monthly variations of the average wind speed. (c) The daily variation of wind speed in February. (d) The daily variation of wind speed in August. Note: Significant differences between the means of WPS and CS were determined using Student's t-test (*** $p < 0.001$).

3.2. Seasonal, Monthly, and Daily Variations of the HTCI

3.2.1. Seasonal and Monthly Variations of the HTCI and Human Body Perception

As shown in Figure 5a, the HTCI in both the WPS and CS demonstrated certain seasonal variations. Apart from summer, when the HTCI in the WPS was significantly lower than that in the CS ($p < 0.01$), the HTCI in the WPS was higher than that in the CS in the other seasons ($p < 0.001$). From the seasonal perspective, significant variations of the HTCI in the WPS between all seasons was observed ($p < 0.05$). The largest variation occurred in winter (12.17), which could be classified as a very uncomfortable feeling. The lowest variation occurred in summer (4.34), which could be classified as a very comfortable feeling. The variations in spring (6.38) and autumn (5.49) had a middle rank, which could be classified as a comfortable feeling. In the CS, the largest variation occurred in winter (10.90), which could be classified as a very uncomfortable feeling, the lowest variation occurred in autumn (4.26), which could be classified as a very comfortable feeling, and the variations in spring (5.09) and summer (4.86) had a middle rank, which could be classified as a comfortable feeling. The difference between spring and summer was not substantial, but their difference with winter and autumn was significant ($p < 0.05$).

The monthly HTCI in the WPS and CS exhibited a "W"-shaped trend. The value gradually decreased from January to May, slowly increased from June to July, decreased again from August to September, and then eventually increased. Apart from July and August, when the HTCI in the WPS was significantly lower than that in the CS ($p < 0.001$), the HTCI values in the WPS in all other months were larger than those in the CS. The WPS would have given people a very uncomfortable feeling in December, January, and February, an uncomfortable feeling in March and November, a comfortable feeling in April, July, August, and October, and a very comfortable feeling in May, June, and September (Figure 5b).

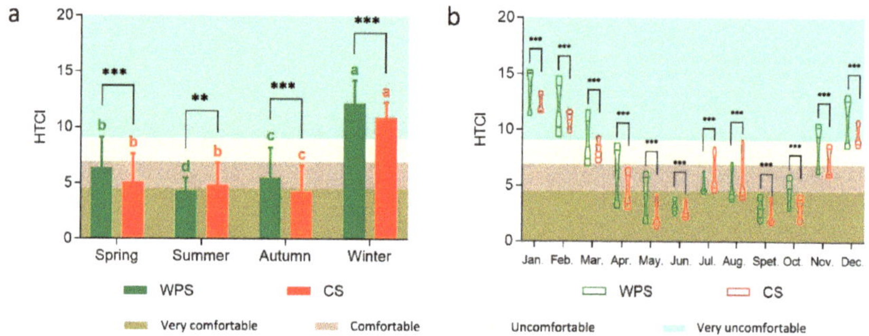

Figure 5. The seasonal and monthly variations of the HTCI in the WPS and CS. (**a**) The seasonal variation of the HTCI; different lowercase letters of the same color indicate results of a one-way ANOVA followed by Tukey's test ($p < 0.05$), and error bars represent the standard deviation. (**b**) The monthly variations of the HTCI, and error bars represent the standard deviation. Note: Significant differences between the means of WPS and CS were determined using Student's *t*-test (** $p < 0.01$, *** $p < 0.001$).

3.2.2. Daily Variation of the HTCI

As shown in Figure 6, the HTCI displayed different trends of daily variation over the 12-month observation period.

In January, February, March, April, October, November, and December, the HTCI in both the WPS and CS exhibited an increasing trend, then a decreasing trend, and finally rebounded. Two relationships existed between the WPS and CS in terms of the magnitude of the HTCI. Take February as an example. The HTCI in the WPS was lower than that in the CS during the period of 9:45–17:00, while it was higher than that in the CS in other periods (Figure 6b). The same relationship occurred in January, March, April, and December. In contrast, in October and November, the daily HTCI values in the WPS were higher than those in the CS in all time periods.

In May, the HTCI in the WPS exhibited a trend of increasing, decreasing, then increasing again, whereas the HTCI in the CS exhibited a "W"-shaped trend. During 12:45–15:30, the HTCI in the WPS was lower than that in the CS, whereas it was higher than that in the CS in other periods (Figure 6e).

In June and September, the HTCI in the WPS exhibited a "W"-shaped trend, while the HTCI in the CS presented a trend of decreasing followed by increasing and reached its peak at around 14:00, after which it slowly decreased (Figure 6f,i). The HTCI in the WPS was lower than that in the CS during the daytime. For example, in June, the HTCI in the WPS was lower than that in the CS during 9:45–20:00, but was higher than that in the CS in other periods (Figure 6f).

In July and August, the HTCI in both the WPS and CS exhibited a trend of increasing then decreasing, and reached its peak at around 14:00, whereas the HTCI in the WPS was lower than that in the CS in almost all time periods (Figure 6g,h).

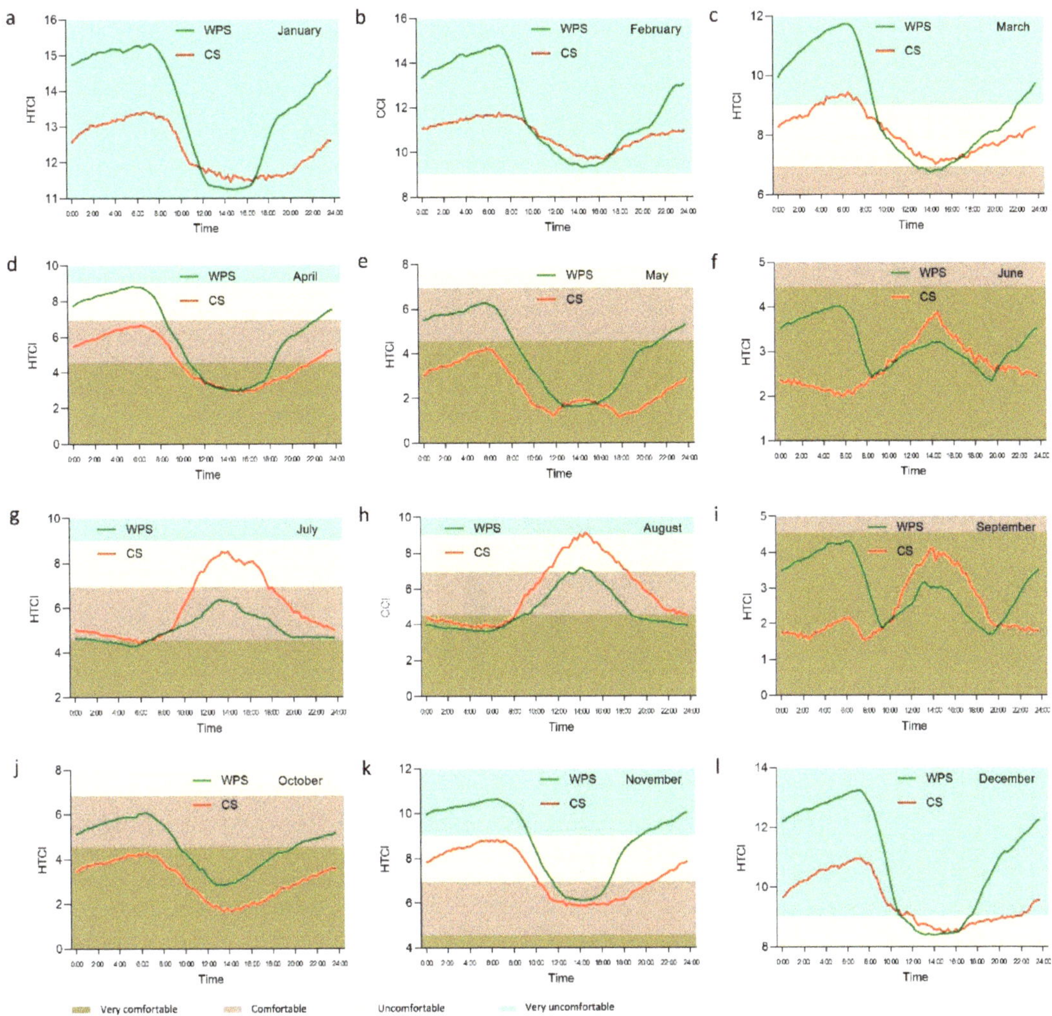

Figure 6. The daily variation of the HTCI between (**a**) January and (**l**) December in the WPS and CS.

3.2.3. Daily Variation of the Human Body Perception

As shown in Figure 6, in January, February, and December, the WPS and CS generated very uncomfortable or uncomfortable feelings all day long (Figure 6a,b,l). In March, the CS generated a very uncomfortable or uncomfortable feeling all day long, while the WPS generated a comfortable feeling during 12:45–15:30 but a very uncomfortable or uncomfortable feeling in other periods (Figure 6c). In April, the WPS generated a comfortable or very comfortable feeling during 8:30–22:15, but an uncomfortable feeling in other periods, while the CS generated a comfortable or very comfortable feeling all day long (Figure 6d). In July, the WPS generated a comfortable or very comfortable feeling all day long, while the CS generated an uncomfortable feeling during 11:00–17:30 but a comfortable or very comfortable feeling in other periods (Figure 6g). In August, the WPS generated an uncomfortable feeling during 13:45–15:00, but a comfortable or very comfortable feeling in other periods, while the CS generated a very uncomfortable or uncomfortable feeling during 11:00–18:15, but a comfortable or very comfortable feeling in other periods (Figure 6h). In November, the WPS generated a comfortable feeling during 11:45–16:30, but a very uncomfortable or uncomfortable feeling in other periods, while the CS generated a comfortable feeling during 10:15–20:30, but an uncomfortable feeling in other periods (Figure 6k). In May, June, September, and October, the WPS and CS both generated a comfortable or very comfortable feeling all day long (Figure 6e,f,i,j).

3.3. Seasonal, Monthly, and Daily Variations of the CTI

3.3.1. Seasonal and Monthly Variations of the CTI and Clothing Suggestions

As shown in Figure 7a, the CTI values in the WPS were all higher than those in the CS in all four seasons ($p < 0.01$). The CTI in the WPS and CS exhibited significant seasonal variations ($p < 0.05$), with winter displaying the highest values (17.14 and 15.49). This belongs to level IV, so a sweater in addition to a thin coat or thin cotton coat is recommended for outdoor activities. Summer had the lowest variations (3.21 and 1.94). This belongs to level II, so a long-sleeve shirt is recommended for outdoor activities. Spring (10.50 and 9.32) and autumn (9.03 and 7.93) ranked in the middle. These belong to level III, so a shirt in addition to a jacket or suit is recommended for outdoor activities.

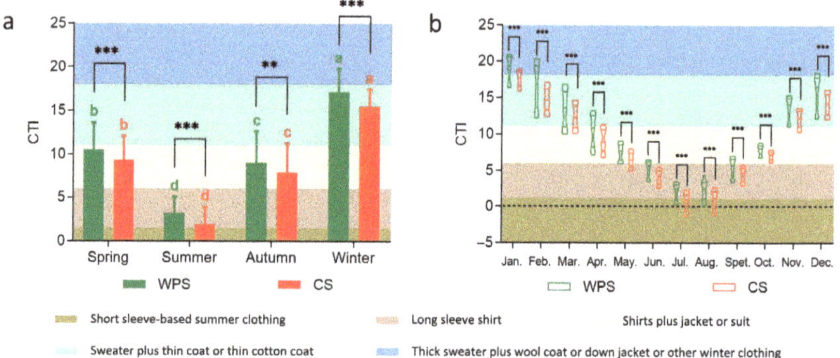

Figure 7. The seasonal and monthly variations of the CTI in the WPS and CS. (**a**) The seasonal variations of the CTI; different lowercase letters of the same color indicate results of a one-way ANOVA followed by Tukey's test ($p < 0.05$), and error bars represent the standard deviation. (**b**) The monthly variations of the CTI. Significant differences between the means of WPS and CS were determined using Student's *t*-test (** $p < 0.01$, *** $p < 0.001$).

From the monthly perspective, the CTI in the WPS and CS demonstrated a decreasing and then increasing trend, with July having the lowest variation (-2.06 and 0.68). The CTI values in the WPS were all higher than those in the CS in each month ($p < 0.01$). In January, the CTI in the WPS belonged to level V, so a thick sweater in addition to a wool coat or down jacket or other winter clothing is recommended for outdoor activities. In the CS, the CTI belonged to level IV, so a sweater in addition to a thin coat or thin cotton coat is recommended for outdoor activities. In February, March, November, and December, the CTI values in the WPS and CS belonged to level IV, so a sweater in addition to a thin coat or thin cotton coat is recommended for outdoor activities. In April, May, and October, the CTI values in the WPS and CS belonged to level III, so a shirt in addition to a jacket or suit is recommended for outdoor activities. In June and September, the CTI values in the WPS and CS belonged to level II, and a long-sleeve shirt is recommended for outdoor activities. In July and August, the CTI values in the WPS belonged to level II, and a long-sleeve shirt is recommended for outdoor activities. In the CS, the CTI values belonged to level I, so short-sleeved summer clothing is recommended for outdoor activities (Figure 7b).

3.3.2. Daily Variation of the CTI and Clothing Suggestions

As shown in Figure 8, the daily variations of the CTI in the WPS and CS in each month demonstrated an increasing and then decreasing trend, followed by a slowly increasing trend. In February, March, and December, the CTI values in the WPS were larger than those in the CS in certain time periods (Figure 8b,c,l). Taking March for example, the CTI in the WPS was lower than that in the CS during the period of 10:15–15:30, whereas it was higher in other periods (Figure 8c). Apart from February, March, and December, the CTI values in the WPS were higher than those in the CS in all periods in all other months (Figure 8a,d–k).

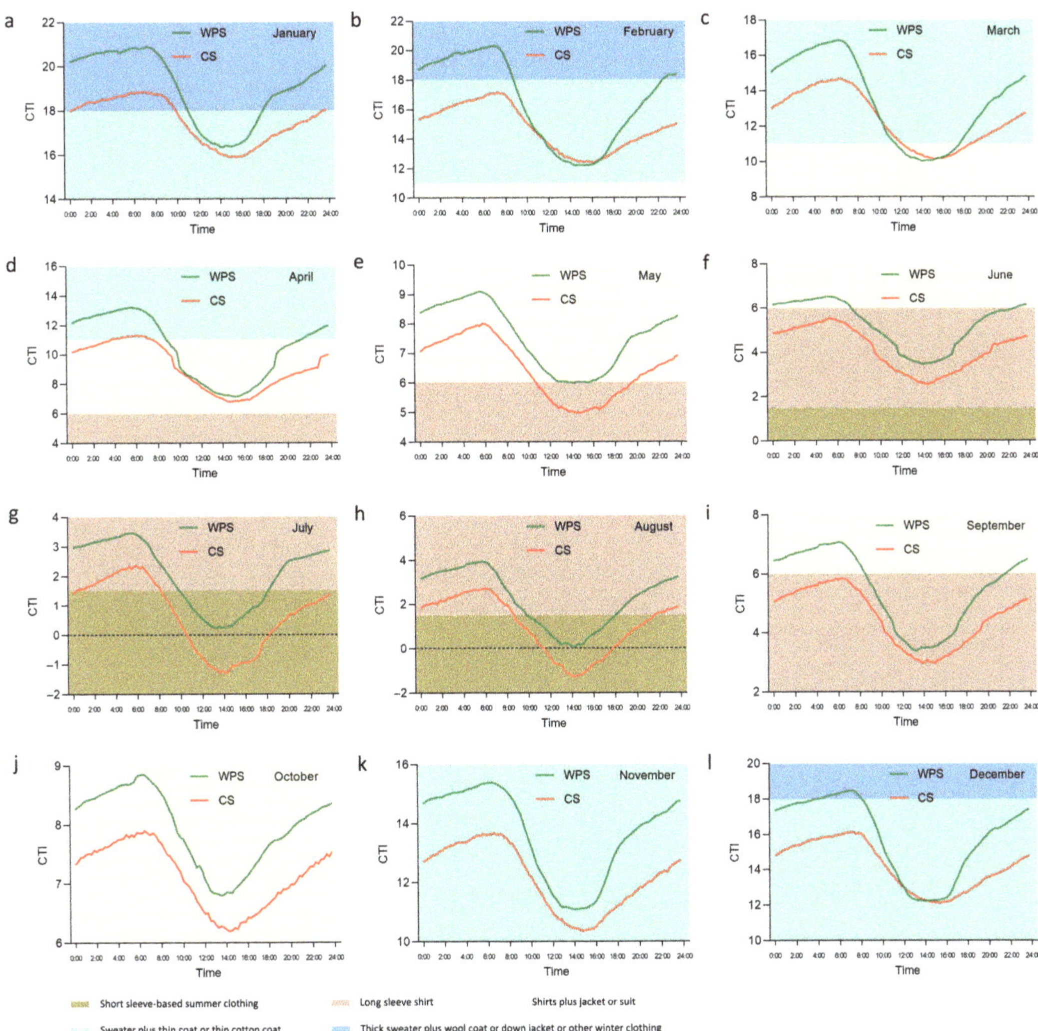

Figure 8. The daily variation of the CTI between (**a**) January and (**l**) December in the WPS and CS.

Clothing suggestions are further discussed based on the daily variation of the CTI. In January, the daily variation of the CTI in both the WPS and CS was reduced from level V to level IV before increasing to level V (Figure 8a). In February, the daily variation of the CTI in the WPS reduced from level V to level IV before increasing to level V, while that in the CS stayed in level IV over the whole day (Figure 8b). In March, the daily variation of the CTI in both the WPS and CS was reduced from level IV to level III before increasing to level IV (Figure 8c). In April, the daily variation in the WPS was reduced from level IV to level III before increasing to level IV, while that in the CS increased from level III to level IV before reducing to level III (Figure 8d). In May, the WPS stayed at level III over the whole day, while that in the CS was reduced from level III to level II before increasing to level III (Figure 8e). In June and September, the daily variation of the CTI in the WPS was reduced from level III to level II before increasing to level III, while that in the CS stayed at level II over the whole day (Figure 8f,i). In July, the daily variation of the CTI in the WPS was reduced from level II to level I before increasing to level II, while that in the CS increased from level I to level II before reducing to level I (Figure 8g). In August, the daily variation of the CTI in both the WPS and CS was reduced from level II to level I before increasing to level II (Figure 8h). In October, both the WPS and CS stayed at level III over the whole day (Figure 8j). In November, both the WPS and CS stayed at level IV over the whole day (Figure 8k). Finally, in December, the daily variation of the CTI in the WPS increased from level IV to level V before reducing to level IV, while the CS stayed at level IV over the whole day (Figure 8l).

4. Discussion

4.1. Ecosystem Services of Urban Wetland Parks in Terms of Microclimate Improvement

This study explored the microclimate and HTC of urban wetland parks, and the seasonal, monthly, and daily variations of the microclimate, HTCI, and CTI were quantitatively analyzed. The results demonstrate that the WPS displayed evident seasonal and monthly changes in the microclimate, suggesting that the microclimates of urban wetland parks are affected primarily by the overall water and heat conditions and maintain similar changes as the regional climate [44]. HTCI was studied on an annual scale, which resulted in a relatively larger range fluctuation. This phenomenon of large ranges on a larger space–time scale has been mentioned in other studies by Vinogradova (2021) [45] and An et al. (2021) [46]. Therefore, in order to obtain a perception of a certain scenario, we also mainly studied the monthly and daily variations of HTCI. The ecological benefits exhibited reduced temperature, increased humidity, and decreased wind speed in summer and other hot months, which confirms that urban wetland parks could provide good ecological services in terms of mitigating the heat island effect and dry island effect.

The findings have been supported by other studies [16,47–50]. The main reason for these is that, compared with gray infrastructure (impermeable concrete and buildings), green infrastructure, especially trees and water bodies, can absorb solar radiation, reduce the land surface temperature, and reduce the increase in vapor caused by solar radiation [17,51,52]. Moreover, ecological infrastructures that are reliant on vegetation can reduce the temperature via shading and increase heat fluxes, because they prevent the solar radiation from reaching the surface and through evapotranspiration to form low-temperature areas under canopies or in grasslands) [18,53]. Additionally, the natural branching configuration of plants can directly block the wind, produce a wind barrier effect, and can reduce the wind energy via the swinging of branches, thereby reducing wind speed (see Figure 9).

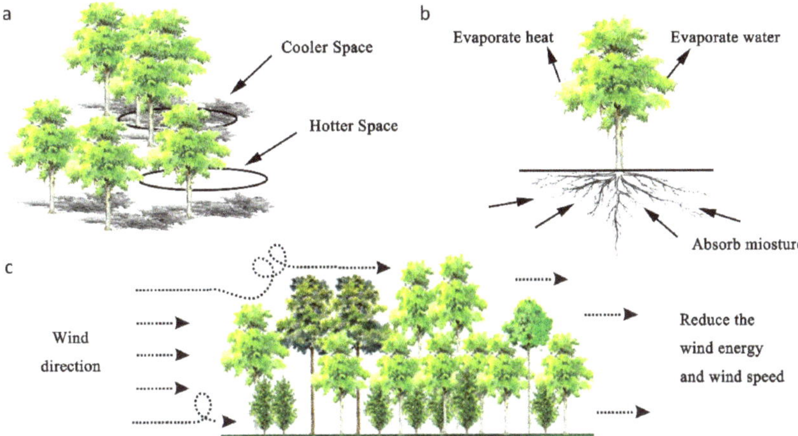

Figure 9. Schematic diagrams of the (**a**) shading, (**b**) transpiration, and (**c**) wind barrier effects of plant communities.

Some interesting findings based on the further analysis of the daily variation in each month were also presented. It was found that urban wetland parks have a warming effect in the midday of the cold months, while they can effectively reduce fluctuations in the daily variation of the wind speed in warmer months. The lack of vegetation and the thermal conductivity of large, impervious surfaces in urban environments result in faster heat loss during colder months [54–56]. Additionally, urban forests will also modify the airflow, thereby causing local strong winds, and the wind chill benefit is obvious during colder months. However, the ecological infrastructures in urban wetland parks have complex spatial structures, and therefore do not have the smooth, reflective planes of artificial facilities. Thus, when solar radiation reaches living organisms such as plants, it can be trapped by plants or diffuse, thereby affecting the energy exchange between vegetation patches [57] (see Figure 10). Moreover, the ecological infrastructures in urban wetland parks also offset a portion of the wind energy via branches and leaves, thereby reducing the daily variation of the wind speed. Therefore, urban wetland parks exert a certain heat preservation effect in the midday period of cold months, rather than a real warming effect.

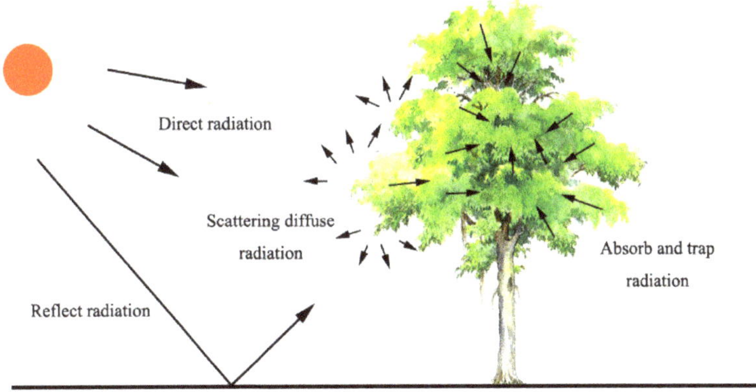

Figure 10. A diagram of the absorption and diffuse reflection of solar radiation by plants.

4.2. HTC Features of Urban Wetland Parks

HTC is a parameter based on temperature, relative humidity, and wind. The data collected and analyzed in this study indicate that the HTCI in the WPS presented significant seasonal variation. In spring, summer, and autumn, the HTCI in the WPS was at a very comfortable and comfortable level, respectively, whereas in winter, it was at a very uncomfortable level. However, the urban wetland park was found to have a better comfort level than the urban environment only in summer (July and August).

The daily variation of the HTC in each month was also discussed. Xixi National Wetland Park can provide a comfortable environment throughout the day in May, June, July, September, and October, as well as at midday in March, April, and November. A comfortable environment was not available at noon in August or throughout the day in winter (December, January, and February). However, compared with the urban environment, Xixi National Wetland Park was found to have significantly improved comfort in the midday hours of winter.

For Hangzhou, China, which has a subtropical monsoon climate, compared with other climatic factors, humidity has a greater impact on the comfort perception of the human body throughout the year. Particularly, in winter, the "cold and wet" climate features are very unfavorable to people staying outdoors [58]. In the cooler months, urban wetland parks can increase outdoor comfort in the warm midday hours. In August, compared with the urban environment, the urban wetland park was found to provide a better comfort level in the midday hours, but the high temperature and high humidity are not sufficient to make a person feel comfortable. Hence, based on the comparison between the comfort levels provided by the urban wetland park and urban environment, in an urban wetland park with a subtropical monsoon climate, July and August provide the largest improvements of HTC.

4.3. Clothing Suggestions

According to the definition and purpose of the CTI, although Xixi National Wetland Park is in a subtropical climate area, the residents are suggested to wear heavy winter clothes when they go out in the cold season due to the restriction of high air humidity, especially at night. Recreational activities with accommodation are not recommended in winter, and specific protective measures should be taken to make visitors feel comfortable.

When visitors go to Xixi National Wetland Park in summer, they will feel much more comfortable wearing cool summer clothes. In August, they are suggested to take good protective measures (e.g., to wear a sunshade hat and sun-protective clothing) and enjoy their activities in the morning, evening, or at night, not at midday.

In this study, the HTCI and CTI were synthetically calculated by climate factors. While both indexes have had numerous applications since their conception, they both indirectly reflect the human body perception and provide clothing suggestions. Hence, in future research, determining the direct perceptions of volunteers and further combining these data with climate data to carry out an evidence-based study will be conducive to research on the relationship between the ecological environment and human health.

5. Conclusions

The increase of gray infrastructure (e.g., concrete buildings, hard pavements, and metal materials) and the decrease of ecological infrastructure (e.g., greenbelts, wetlands, and water bodies) change the underlying structure of the urban ecological environment, thereby affecting the ecosystem services in urban areas [52]. In this study, the effect of urban wetland parks on HTC over one year was quantified, and the results indicate that urban wetland parks can mitigate the heat island effect and dry island effect (by reducing the temperature, increasing the humidity, and reducing the wind speed) in summer, thereby exhibiting a good ecological function. More importantly, urban wetland parks can provide ecological services at midday during winter to mitigate the cold island effect, thereby exerting a certain heat preservation effect. Additionally, urban wetland parks were found to exhibit their best performance in improving HTC during the daytime of the hot season (June, July, August, and September, and especially the whole day in July and August) and the midday period of the cold season (December, January, February, and March). However, improvements in other months (especially in October and November) were not authenticated by the data analyzed in this study. Finally, based on the findings of this study, it is suggested that citizens should take good protective measures and enjoy their activities in the morning, evening, or at night, not at midday in hot weather. Moreover, extra layers are suggested to be worn before visiting urban wetland parks at night in cold weather, and recreational activities involving accommodation are not recommended.

In urban planning, more green space (plants) and blue space (water bodies) should be introduced, and the effect of impervious surfaces on the land surface climate should be reduced to create a microclimate conducive to human health. Administrators and policymakers should consider detailed management and strategies in parks and should plan indoor and outdoor activities for visitors to induce the most comfort and relaxation.

Finally, the HTCI is based upon a thermal stress index that does not account for radiation, which is difficult to comprehensively characterize from a microclimatic perspective. Meanwhile, like other empirical indexes, the HTCI is also intrinsically unable to take into account metabolic rate and clothing insulation. These are the certain limitations of this study. In addition, because each thermal index has its own applicable scope, it is suggested that, in addition to focusing on the research object and its thermal environment condition, the selection of the index and the formulation of a monitoring program should take into account the clothing insulation effect, metabolic activity changes, and the spatial heterogeneity of temperature [37].

Supplementary Materials: The following are available online at https://www.mdpi.com/article/10.3390/f12101322/s1, Figure S1: The daily variations of the average temperature, Figure S2: The daily variations of the average relative humidity, Figure S3: The daily variations of the average wind speed.

Author Contributions: Conceptualization, Z.Z. and B.Y.; writing—original draft preparation, Z.Z. and J.D.; writing—review & editing, Z.Z. and Q.H.; supervision, B.Y. All authors have read and agreed to the published version of the manuscript.

Funding: This research is supported by the Fundamental Research Funds for CAF (CAFYBB2019ZC008).

Institutional Review Board Statement: Not applicable.

Informed Consent Statement: Not applicable.

Data Availability Statement: Data is contained within the article and Supplementary Materials.

Conflicts of Interest: The authors declare no conflict of interest.

References

1. Lai, D.; Zhou, C.; Huang, J.; Jiang, Y.; Long, Z.; Chen, Q. Outdoor space quality: A field study in an urban residential community in central China. *Energy Build.* **2014**, *68*, 713–720. [CrossRef]
2. Schinasi, L.H.; Benmarhnia, T.; De Roos, A.J. Modification of the association between high ambient temperature and health by urban microclimate indicators: A systematic review and meta-analysis. *Environ. Res.* **2018**, *161*, 168–180. [CrossRef]
3. Tsavdaroglou, M.; Al-Jibouri, S.H.; Bles, T.; Halman, J.I. Proposed methodology for risk analysis of interdependent critical infrastructures to extreme weather events. *Int. J. Crit. Infrastruct. Prot.* **2018**, *21*, 57–71. [CrossRef]
4. Wazneh, H.; Arain, M.A.; Coulibaly, P. Climate indices to characterize climatic changes across southern Canada. *Meteorol. Appl.* **2019**, *27*, e1861. [CrossRef]
5. Mitchell, M.G.E.; Suarez-Castro, A.F.; Martinez-Harms, M.; Maron, M.; McAlpine, C.; Gaston, K.J.; Johansen, K.; Rhodes, J.R. Reframing landscape fragmentation's effects on ecosystem services. *Trends Ecol. Evol.* **2015**, *30*, 190–198. [CrossRef] [PubMed]
6. Wlemm, K.; Heusinkveld, B.G.; Lenzholzer, S.; Hove, B. Street greenery and its physical and psychological impact on thermal comfort. *Landsc. Urban Plan.* **2015**, *138*, 87–98.
7. Mei, Y.; Sohngen, B.; Babb, T. Valuing urban wetland quality with hedonic price model. *Ecol. Indic.* **2018**, *84*, 535–545. [CrossRef]
8. Bertassello, L.E.; Rao, P.S.C.; Park, J.; Jawitz, J.W.; Botter, G. Stochastic modeling of wetland-groundwater systems. *Adv. Water Resour.* **2018**, *112*, 214–223. [CrossRef]
9. Park, J.; Wang, D.; Kumar, M. Spatial and temporal variations in the groundwater contributing areas of inland wetlands. *Hydrol. Process.* **2019**, *34*, 1117–1130. [CrossRef]

10. Gutzwiller, K.J.; Flather, C. Wetland features and landscape context predict the risk of wetland habitat loss. *Ecol. Appl.* **2011**, *21*, 968–982. [CrossRef]

11. Herbert, E.R.; Boon, P.; Burgin, A.J.; Neubauer, S.; Franklin, R.; Ardon, M.; Hopfensperger, K.N.; Lamers, L.P.M.; Gell, P. A global perspective on wetland salinization: Ecological consequences of a growing threat to freshwater wetlands. *Ecosphere* **2015**, *6*, art206. [CrossRef]

12. Li, N.; Tian, X.; Li, Y.; Fu, H.; Jia, X.; Jin, G.; Jiang, M. Seasonal and Spatial Variability of Water Quality and Nutrient Removal Efficiency of Restored Wetland: A Case Study in Fujin National Wetland Park, China. *Chin. Geogr. Sci.* **2018**, *28*, 1027–1037. [CrossRef]

13. Vanos, J.K.; Warland, J.S.; Gillespie, T.J.; Kenny, N.A. Review of the physiology of human thermal comfort while exercising in urban landscapes and implications for bioclimatic design. *Int. J. Biometeorol.* **2010**, *54*, 319–334. [CrossRef] [PubMed]

14. Cheng, X.; Yang, L.; Cao, J.; Liang, D.; Song, G. Knowledge Mapping on Literature Research Progress of Human Thermal Comfort in buildings Based on Web of Science. *IOP Conf. Ser. Earth Environ. Sci.* **2020**, *514*. [CrossRef]

15. Blazejczyk, K.; Epstein, Y.; Jendritzky, G.; Staiger, H.; Tinz, B. Comparison of UTCI to selected thermal indices. *Int. J. Biometeorol.* **2011**, *56*, 515–535. [CrossRef] [PubMed]

16. Mughal, M.O.; Kubilay, A.; Fatichi, S.; Meili, N.; Carmeliet, J.; Edwards, P.; Burlando, P. Detailed investigation of microclimate by means of computational fluid dynamics (CFD) in a tropical urban environment. *Urban Clim.* **2021**, *39*, 10093. [CrossRef]

17. Meili, N.; Manoli, G.; Burlando, P.; Carmeliet, J.; Chow, W.T.L.; Coutts, A.M.; Roth, M.; Velasco, E.; Vivoni, E.R.; Fatichi, S. Tree effects on urban microclimate: Diurnal, seasonal, and climatic temperature differences explained by separating radiation, evapotranspiration, and roughness effects. *Urban For. Urban Green.* **2021**, *58*, 126970. [CrossRef]

18. Meili, N.; Acero, J.A.; Peleg, N.; Manoli, G.; Burlando, P.; Fatichi, S. Vegetation cover and plant-trait effects on outdoor thermal comfort in a tropical city. *Build. Environ.* **2021**, *195*, 107733. [CrossRef]

19. Lai, D.; Guo, D.; Hou, Y.; Lin, C.; Chen, Q. Studies of outdoor thermal comfort in northern China. *Build. Environ.* **2014**, *77*, 110–118. [CrossRef]

20. Chatzidimitriou, A.; Yannas, S. Microclimate development in open urban spaces: The influence of form and materials. *Energy Build.* **2015**, *108*, 156–174. [CrossRef]

21. Estoque, R.C.; Murayama, Y.; Myint, S.W. Effects of landscape composition and pattern on land surface temperature: An urban heat island study in the megacities of Southeast Asia. *Sci. Total. Environ.* **2017**, *577*, 349–359. [CrossRef] [PubMed]

22. Yang, B.; Olofsson, T.; Nair, G.; Kabanshi, A. Outdoor thermal comfort under subarctic climate of north Sweden—A pilot study in Umeå. *Sustain. Cities Soc.* **2017**, *28*, 387–397. [CrossRef]

23. Zhao, Q.; Sailor, D.J.; Wentz, E.A. Impact of tree locations and arrangements on outdoor microclimates and human thermal comfort in an urban residential environment. *Urban For. Urban Green.* **2018**, *32*, 81–91. [CrossRef]

24. Pan, L.; Cui, L.; Wu, M. Tourist behaviors in wetland park: A preliminary study in Xixi National Wetland Park, Hangzhou, China. *Chin. Geogr. Sci.* **2010**, *20*, 66–73. [CrossRef]

25. Steadman, R.G. The assessment of sultriness. Part I: A temperature-humidity index based on human physiology and clothing science. *J. Appl. Meteorol.* **1979**, *18*, 861–873. [CrossRef]

26. Sirangelo, B.; Caloiero, T.; Coscarelli, R.; Ferrari, E.; Fusto, F. Combining stochastic models of air temperature and vapour pressure for the analysis of the bioclimatic comfort through the Humidex. *Sci. Rep.* **2020**, *10*, 11395. [CrossRef]

27. Masterton, J.M.; Richardson, F.A. *Humidex: A Method of Quantifying Human Discomfort Due to Excessive Heat and Humidity*; CLI1-79, Environment Canada, Atmospheric Environment: Toronto, ON, Canada, 1979.

28. Houghton, F.C.; Yaglou, C.P. Determining Equal Comfort Lines. *J. Am. Soc. Heat. Vent. Eng.* **1923**, *29*, 165–176.

29. ISO 7243. *Hot environments—Estimation of the Heat Stress on Working Man, Based on the WBGT Index (Wet Bulb Globe Temperature)*; International Standardization Organisation: Geneva, Switzerland, 1989.

30. Alfano, F.R.D.; Palella, B.I.; Riccio, G. On the Problems Related to Natural Wet Bulb Temperature Indirect Evaluation for the Assessment of Hot Thermal Environments by Means of WBGT. *Ann. Occup. Hyg.* **2012**, *56*, 1063–1079. [CrossRef]

31. Lin, L.; Luo, M.; Chan, T.O.; Ge, E.; Liu, X.; Zhao, Y.; Liao, W. Effects of urbanization on winter wind chill conditions over china. *Sci. Total Environ.* **2019**, *688*, 389–397. [CrossRef] [PubMed]

32. Gagge, A.P.; Stolwijk, J.A.J.; Nishi, Y. An effective temperature scale based on a simple model of human physiological regulatory response. *Ashrae Trans.* **1971**, *77*, 21–36.

33. Gagge, A.P.; Fobelets, A.P.; Berglund, L.G. A standard predictive index of human response to the thermal environment. *Ashrae Trans.* **1986**, *92*, 709–731.

34. ISO 11079. *Evaluation of Cold Environments: Determination of Required Clothing Insulation (IREQ)*; International Standardization Organisation: Geneva, Switzerland, 2007.

35. Alfano, F.R.D.; Palella, B.I.; Riccio, G. Notes on the implementation of the IREQ model for the assessment of extreme cold environments. *Ergonomics* **2013**, *56*, 707–724. [CrossRef] [PubMed]

36. ISO 7933. *Ergonomics of the Thermal Environment: Analytical Determination and Interpretation of Heat Stress Using Calculation of the Predicted Heat Strain*; International Standardization Organisation: Geneva, Switzerland, 2004.

37. Alfano, F.R.D.; Palella, B.I.; Riccio, G. Thermal Environment Assessment Reliability Using Temperature—Humidity Indices. *Ind. Health* **2011**, *49*, 95–106. [CrossRef] [PubMed]

38. Höppe, P. The physiological equivalent temperature—A universal index for the biometeorological assessment of the thermal environment. *Int. J. Biometeorol.* **1999**, *43*, 71–75. [CrossRef] [PubMed]
39. Jendritzky, G.; Bessemoulin, P.; Blazejczyk, K.; Cegnar, T.; Dittmann, E.; Fiala, D.; Nicol, F.; Havenith, G.; Hassi, J.; Höppe, P.; et al. *Towards a Universal Thermal Climate Index UTCI for Assessing the Thermal Environment of the Human Being*; Final Report COST Action 730; European Concerted Research Action: Freiburg, Germany, 2009.
40. Lu, D.; Cui, S.; Li, C. *The Influence of Beijing Urban Greening and Summer Microclimate Conditions on Human Fitness*; Forestry and Metrology Papers; Meteorological Press: Beijing, China, 1984; pp. 144–152. (In Chinese)
41. Gu, L.; Wang, C.; Wang, Y.; Wang, X.; Sun, Z.; Wang, Q.; Sun, R. Patterns of temporal variation of microclimate and extent of human comfort in the recreation forests in Huishan National Forest Park. *Sci. Silvae Sin.* **2019**, *55*, 150–159. (In Chinese)
42. Jin, G.; Liu, H.; Lu, Z.; Wu, J.; Xu, L.; Sun, G.; Li, P.; Li, J.; Xu, C. Canopy structures and degree of comfort with urban forests of Beijing in summer. *J. Zhejiang AF Univ.* **2019**, *36*, 550–556. (In Chinese)
43. Zhu, L.; Qian, P.; Qian, Y. Weather index study for clothing. *Sci. Meteorol. Sin.* **2001**, *21*, 468–473. (In Chinese)
44. Rapp, J.M.; Silman, M.R. Diurnal, seasonal, and altitudinal trends in microclimate across a tropical montane cloud forest. *Clim. Res.* **2012**, *55*, 17–32. [CrossRef]
45. Vinogradova, V. Using the Universal Thermal Climate Index (UTCI) for the assessment of bioclimatic conditions in Russia. *Int. J. Biometeorol.* **2020**, *65*, 1473–1483. [CrossRef]
46. An, L.; Hong, B.; Cui, X.; Geng, Y.; Ma, X. Outdoor thermal comfort during winter in China's cold regions: A comparative study. *Sci. Total Environ.* **2021**, *768*, 144464. [CrossRef]
47. Georgi, N.J.; Zafiriadis, K. The impact of park trees on microclimate in urban areas. *Urban Ecosyst.* **2006**, *9*, 195–209. [CrossRef]
48. Wu, Z.; Chen, L. Optimizing the spatial arrangement of trees in residential neighborhoods for better cooling effects: Integrating modeling with in-situ measurements. *Landsc. Urban Plan.* **2017**, *167*, 463–472. [CrossRef]
49. Atwa, S.; Ibrahim, M.G.; Murata, R. Evaluation of plantation design methodology to improve the human thermal comfort in hot-arid climatic responsive open spaces. *Sustain. Cities Soc.* **2020**, *59*, 102198. [CrossRef]
50. Hu, L.; Li, Q. Greenspace, bluespace, and their interactive influence on urban thermal environments. *Environ. Res. Lett.* **2020**, *15*, 034041. [CrossRef]
51. Lin, Y.-H.; Tsai, K.-T. Screening of Tree Species for Improving Outdoor Human Thermal Comfort in a Taiwanese City. *Sustainability* **2017**, *9*, 340. [CrossRef]
52. Soydan, O. Effects of landscape composition and patterns on land surface temperature: Urban heat island case study for Nigde, Turkey. *Urban Clim.* **2020**, *34*, 100688. [CrossRef]
53. Gkatsopoulos, P. A Methodology for Calculating Cooling from Vegetation Evapotranspiration for Use in Urban Space Microclimate Simulations. *Procedia Environ. Sci.* **2017**, *38*, 477–484. [CrossRef]
54. Liu, W.; Zhang, Y.; Deng, Q. The effects of urban microclimate on outdoor thermal sensation and neutral temperature in hot-summer and cold-winter climate. *Energy Build.* **2016**, *128*, 190–197. [CrossRef]
55. Kántor, N.; Kovács, A.; Takács, Á. Seasonal differences in the subjective assessment of outdoor thermal conditions and the impact of analysis techniques on the obtained results. *Int. J. Biometeorol.* **2016**, *60*, 1615–1635. [CrossRef] [PubMed]
56. Yuan, C.; Adelia, A.S.; Mei, S.; He, W.; Li, X.-X.; Norford, L. Mitigating intensity of urban heat island by better understanding on urban morphology and anthropogenic heat dispersion. *Build. Environ.* **2020**, *176*, 106876. [CrossRef]
57. Gaudio, N.; Gendre, X.; Saudreau, M.; Seigner, V.; Balandier, P. Impact of tree canopy on thermal and radiative microclimates in a mixed temperate forest: A new statistical method to analyse hourly temporal dynamics. *Agric. For. Meteorol.* **2017**, *237–238*, 71–79. [CrossRef]
58. Peng, X.; Wu, W.; Zheng, Y.; Sun, J.; Hu, T.; Wang, P. Correlation analysis of land surface temperature and topographic elements in Hangzhou, China. *Sci. Rep.* **2020**, *10*, 1–16. [CrossRef] [PubMed]

Article

The Nature and Size Fractions of Particulate Matter Deposited on Leaves of Four Tree Species in Beijing, China

Huixia Wang [1], Yan Xing [2], Jia Yang [1], Binze Xie [1], Hui Shi [1,*] and Yanhui Wang [3,*]

[1] School of Environmental and Municipal Engineering, Xi'an University of Architecture and Technology, Xi'an 710055, China; wanghuixia@xauat.edu.cn (H.W.); yangjia1802@126.com (J.Y.); xiebinzejianda@126.com (B.X.)

[2] Shaanxi Environmental Monitoring Centre, Xi'an 710054, China; xy18792833799@126.com

[3] Institute of Forest Ecology, Environment and Nature Protection, Chinese Academy of Forestry, Beijing 100091, China

* Correspondence: shihui@xauat.edu.cn (H.S.); wangyh@caf.ac.cn (Y.W.)

Abstract: Particulate matter (PM) in different size fractions ($PM_{0.1–2.5}$, $PM_{2.5–10}$ and $PM_{>10}$) accumulation on four tree species (*Populus tomentosa*, *Platanus acerifolia*, *Fraxinus chinensis*, and *Ginkgo biloba*) at two sites with different pollution levels was examined in Beijing, China. Among the tested tree species, *P. acerifolia* was the most efficient species in capturing PM, followed by *F. chinensis*, *G. biloba*, and *P. tomentosa*. The heavily polluted site had higher PM accumulation on foliage and a higher percentage of $PM_{0.1–2.5}$ and $PM_{2.5–10}$. Encapsulation of PM within cuticles was observed on leaves of *F. chinensis* and *G. biloba*, which was further dominated by $PM_{2.5}$. Leaf surface structure explains the considerable differences in PM accumulation among tree species. The amounts of accumulated PM ($PM_{0.1–2.5}$, $PM_{2.5–10}$, and $PM_{>10}$) increased with the increase of stomatal aperture, stomatal width, leaf length, leaf width, and stomatal density, but decreases with contact angle. Considering PM accumulation ability, leaf area index, and tolerance to pollutants in urban areas, we suggest *P. acerifolia* should be used more frequently in urban areas, especially in "hotspots" in city centers (e.g., roads/streets with heavy traffic loads). However, *G. biloba* and *P. tomentosa* should be installed in less polluted areas.

Keywords: air pollution alleviation; accumulation on leaves; $PM_{2.5}$; encapsulated particles; urban trees

Citation: Wang, H.; Xing, Y.; Yang, J.; Xie, B.; Shi, H.; Wang, Y. The Nature and Size Fractions of Particulate Matter Deposited on Leaves of Four Tree Species in Beijing, China. *Forests* **2022**, *13*, 316. https://doi.org/10.3390/f13020316

Academic Editors: Manuel Esperon-Rodrigue and Tina Harrison

Received: 17 January 2022
Accepted: 11 February 2022
Published: 15 February 2022

Publisher's Note: MDPI stays neutral with regard to jurisdictional claims in published maps and institutional affiliations.

1. Introduction

In Beijing, the capital of China, with rapid economic development, urbanization, and industrialization, ecological problems are becoming increasingly prominent. Air pollution, especially particulate matter (PM), has become one of the most severe problems over the past several decades [1]. Studies have shown that PM has recently been ranked fifth among the major risk factors threatening human health globally, and thus is first among environmental risks [2,3]. Particles with a diameter less than 10 μm (PM_{10}) can cause premature mortality, accelerated atherosclerosis, lung cancer, heart disease, asthma, preterm birth, mutagenicity and DNA damage, and inflammatory responses [4–6]. Nevertheless, fine particles (particles with diameter less than 2.5 μm, $PM_{2.5}$) are more toxic and more strongly associated with human health effects than coarse particles (particles with diameter between 2.5 and 10 μm) [7]. Therefore, the air quality standards for PM_{10} and $PM_{2.5}$ in China are set to 75 and 35 μg/m^3 (annual mean), and 150 and 70 μg/m^3 (daily mean), respectively [8]. However, urban Beijing's PM_{10} and $PM_{2.5}$ concentrations are much higher than the national standards [9]. Thus, reducing PM concentrations, especially $PM_{2.5}$, is considered one of the most significant tasks related to environmental protection in urban areas. The government has taken measures to control pollutant sources (e.g., adjusting the industrial structure and promoting energy-saving technology). Meanwhile, massive afforestation is believed to be an additional and helpful measure to alleviate air pollution by filtering and adsorbing PM through forest crown/leaves.

Trees are a significant element in a city's landscape and are the most effective vegetation types with regards to reducing PM [10]. This is mainly due to the fact that trees have extensive leaf areas [11], and the structure of tree crowns changes the turbulence of air movement above and within tree canopies [12]. McDonald et al. [13] estimated that a 3.7–16.5% increase in tree cover in West Midland, UK, could reduce PM_{10} concentrations by 10%. Reductions in total suspended particles (TSP), PM_{10}, PM_7, PM_4, $PM_{2.5}$, and PM_1 associated with an increase in canopy cover have been reported in China, the United States, Chile, and Israel [14–17], indicating a direct positive effect on air quality by removing $PM_{2.5}$ by urban trees. If a massive plantation was employed to decrease airborne $PM_{2.5}$, the differences in $PM_{2.5}$ accumulation among tree species should be considered and the most efficient species should be recognized. Some studies compared the $PM_{2.5}$ capturing ability of different plant species. Räsänen et al. [18] investigated the efficiency (C_p) of *Pinus sylvestris*, *Betula pubescens*, *Tilia vulgaris*, and *Betula pendula* leaves to capture $PM_{2.5}$ using simulation method (i.e., NaCl particles). They found that C_p is influenced by leaf structures (e.g., leaf size, wettability, stomatal density, and leaf hair density). However, simulation studies are different from field investigations. Sæbø et al. [11] compared the PM accumulation on leaves of 47 species, including 22 trees and 25 shrubs, in Norway and Poland. They found a species-related difference in PM accumulation in both countries. The abilities of leaves to accumulate PM and its size fractions could be attributed to leaf morphology (e.g., leaf hair density, leaf roughness, and wax content) [6,19–21].

The detriments of PM on human health are primarily determined by particle size. Thus, the present study examined plants accumulating PM in two aspects. First, the amount of PM and its size fractions ($PM_{0.1–2.5}$, $PM_{2.5–10}$, and $PM_{>10}$) deposited on leaves in two contrasting urban environments were quantified. Second, the influences of anatomical/physiological leaf characteristics (e.g., stomatal density, stomatal size, single leaf area, wettability) on PM (i.e., PM, $PM_{0.1–2.5}$, $PM_{2.5–10}$, and $PM_{>10}$) accumulation abilities were investigated. Four tree species (*Populus tomentosa*, *Platanus acerifolia*, *Fraxinus chinensis*, and *Ginkgo biloba*) were selected as test species due to their prevalence in urban and suburban environments, widespread in temperate regions, and also since they are recommended for extensive plantation in Beijing. The findings of this study can provide the impetus for using urban trees to improve air quality, and provide guidance for the work of urban planners and those involved in environmental protection.

2. Materials and Methods

2.1. Plant Materials and Experimental Sites

Four tree species, *P. tomentosa*, *P. acerifolia*, *F. chinensis*, and *G. biloba*, were selected for this study at Beijing Botanical Garden (Site 1, located in Haidian District, 39°59′29.66″ N, 116°12′40.25″ E, upwind of Beijing) and Huangcun (Site 2, located in Daxing District, 39°42′45.13″ N, 116°19′08.44″ E, downwind of Beijing) (Figure 1). The sampling plants grow in the center of the garden or near a busy road with a traffic density of ~5680 cars/h at Site 1 and Site 2. The two sites showed different $PM_{2.5}$ and PM_{10} concentrations, as measured at the nearest monitoring station operated by the Beijing Municipal Ecological and Environmental Monitoring Center [1].

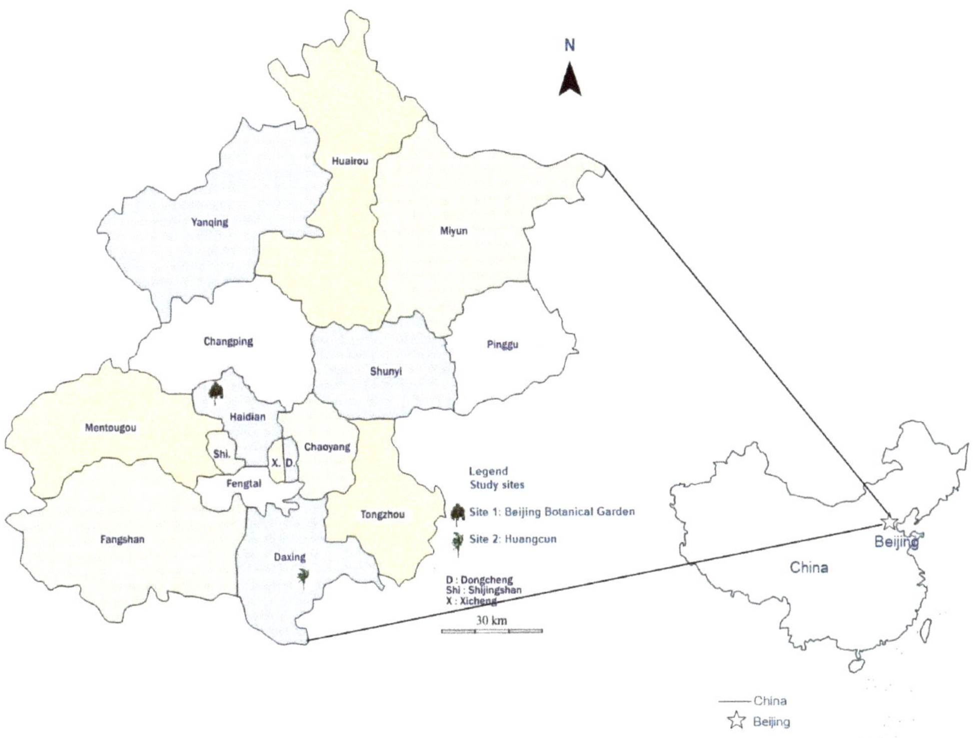

Figure 1. Location of the study sites.

2.2. Sampling Procedure

Leaf sampling was conducted on October 1 and 3, 2014 at Site 1 and Site 2, respectively, when there was no previous rainfall for more than a week. For each species, five individual trees under good growth conditions were selected. At Site 2, the distance between the sampled trees and the road center was about 10 m. Thus, the surrounding environment of each sample tree was similar. Small branches with mature and healthy leaves were cut from four dimensions (N, S, E and W) at 2–6 m above ground at each site and for each species. After cutting, small branches bearing leaves were placed in labeled ziplock bags, transported to the laboratory, and analyzed as soon as possible.

2.3. Analysis of PM

For each plant species at Sites 1 and 2, three batches of leaves were initially prepared. For each batch, 20–30 pieces for *P. acerifolia*, or 30–80 pieces for *P. tomentosa*, *F. chinensis*, and *G. biloba* were selected. The leaves were hand washed using a brush, with 200 mL of ultrapure water (ELGA, High Wycombe, Buckinghamshire, UK). The hemi-surface leaf area was measured using Image J software (Version 1.46; Wayne Rasband, National Institutes of Health, Bethesda, ML, USA) after scanning (HP Scanjet 3570c, HP Inc., Palo Alto, CA, USA). For the filtration procedure, membranes with pore size of 10, 2.5, and 0.1 μm were first soaked in ultrapure water for 2 h and then dried at 105 °C in a drying chamber for 6–8 h to remove soluble impurities. The filters were then put in a balancing chamber for at least 24 h to stabilize. Every membrane was pre-weighed before filtration using a balance with 0.1 mg accuracy (SI-114, Denver Instrument Co., Arvada, CO, USA). The washing solution was hand-shaken for several minutes to re-suspend all washed particles before filtration. Washing solution was then pumped through membranes with pore size of 10, 2.5, and

0.1 µm successively. The filtration was carried out using a 47-mm glass filter funnel with stopper support assembly (Millipore Corp., Bedford, MA, USA) connected to a vacuum pump (SHB-III; Greatwall Scientific Industrial and Trade, Co., Ltd., Zhengzhou, China). Three fractions of PM were collected on the filters: (i) $PM_{>10}$ (particles intercepted by membrane with pore size of 10 µm, large), (ii) $PM_{2.5-10}$ (particles intercepted by membrane with pore size of 2.5 µm, coarse), and (iii) $PM_{0.1-2.5}$ (particles intercepted by membrane with pore size of 0.1 µm, fine). Loaded filters were subsequently dried for more than 24 h at 40°C, stabilized in the weighing room for 30 min, and then re-weighed. Consequently, the pre-weight was subtracted from the post-weight to calculate the mass of PM deposited on leaves in every size fraction of each washed sample. The resulting weight was finally divided by leaf area. At this moment, we obtained the total weight of deposited PM per unit leaf area for each sample, and also for $PM_{0.1-2.5}$, $PM_{2.5-10}$, and $PM_{>10}$.

The deposited $PM_{0.1-2.5}$, $PM_{2.5-10}$, $PM_{>10}$, and PM per unit green land was calculated by multiplying PM per unit leaf area and leaf area index (LAI). The LAI values were 2.13, 3.18, 2.67, and 2.52 for *P. tomentosa*, *P. acerifolia*, *F. chinensis*, and *G. biloba*, respectively, using a LAI-2000 (LI-COR., Inc., Lincoln, NE, USA) at Site 1.

2.4. Analysis of Leaf Surface Characteristics

Field emission scanning electron microscopy (FESEM, Quanta 200 FEG; FEI Company, Hillsboro, OR, USA) was used to determine leaf surface microstructures. Six pieces (three for upper side and three for lower side) of air-dried samples (about 5 mm $\times$ 5 mm) for each species at each site were cut from the center of the leaves, and coated with a thin layer of gold-palladium using a precision etching coating system (Model 682, Ga-tan Co., Ltd., Pleasanton, CA, USA). Stomatal density (per mm^2) was determined by calculating the number of stomata in ten FESEM images with a magnification of $\times 500$. Particle frequency on stomata of each species was counted from 10 FESEM images with a magnification of $\times 1000$. Fifty stomata were randomly selected to measure stomatal length, width, and aperture. Leaf roughness on upper and lower sides was evaluated on a subjective scale (1= relatively smooth, and 5 = very rough) [11] using FESEM images with a magnification of $\times 1000$.

The wettability of leaf surface was evaluated by contact angle (CA) with distilled water. CAs were determined on upper and lower sides at room temperature using a goniometer (Kino SL200A, KINO Industry CO. Ltd., Somerville, Boston, MA, USA). For every species at every site, thirty pieces (about 5 mm $\times$ 5 mm, fifteen for upper side and fifteen for lower side) were cut from the middle of each leaf next to the main vein. Then, these pieces were attached to a glass plate with double-sided tape. A 3-µL water droplet was made with a capillary tube and carefully applied to the leaf surface. A photograph of the profile of each water droplet resting on the leaf surface was taken with a charge-coupled device equipped with a camera within 30 s after placing the water droplet. The digital photographs were downloaded, and CAs were determined using computer software (CAST2.0, KINO Industry CO. Ltd., Somerville, Boston, MA, USA). Then, the mean values were calculated.

Measurements of specific leaf area (SLA) and single leaf area, leaf length, leaf width, and petiole length were made on the same batches as were used for analysis of PM. After scanning, 30 leaves were randomly selected for measurements of single leaf area, leaf length, leaf width and petiole length using Image J software. The batches were dried in a drying chamber at 80 °C for at least 24-h and weighed to produce a value of dry weight. The dry weight and measured leaf area of the batches were used to calculate SLA as cm^2/g (dry weight).

2.5. Data Analysis

Statistical tests were performed with Minitab 16 software (Minitab Ltd., Shanghai, China). One-way analysis of variance was undertaken to estimate the differences in PM accumulation and its size fractions ($PM_{0.1-2.5}$, $PM_{2.5-10}$, and $PM_{>10}$) among different species at each site. The main effects of tree species, different sites, and their interaction on accumulation of PM on leaves and its size fractions ($PM_{0.1-2.5}$, $PM_{2.5-10}$, and $PM_{>10}$) were tested with a two-way analysis of variance. The relative importance of measured leaf characteristics (stomatal density, stomatal length, stomatal width, stomatal aperture, CA upper side, CA lower side, single leaf area, SLA, leaf length, leaf width, petiole length, roughness upper side, roughness lower side) on the amount of PM and its size fractions was evaluated using principal component analysis (PCA) and partial least squares regression. A given effect was assumed to be significant at $p < 0.05$.

3. Results

3.1. Differences in PM and Its Size Fractions among Species

The total PM deposited on leaves was significantly different among four species at each site ($p < 0.001$, Table 1). The amounts of $PM_{0.1-2.5}$, $PM_{2.5-10}$, $PM_{>10}$, and PM ranged from 3.9–14.2, 5.7–41.2, 80.0–109.1, and 89.6–164.5 $\mu g/cm^2$; and 9.3–25.0, 15.9–51.7, 98.7–389.5, and 123.9–466.2 $\mu g/cm^2$, at Site 1 and Site 2, respectively. Among the four species, *P. acerifolia* had the greatest accumulated PM, followed by *F. chinensis*, *G. biloba*, and *P. tomentosa*. The mass of $PM_{0.1-2.5}$, $PM_{2.5-10}$, and $PM_{>10}$ accumulated on leaves also showed significant differences among tree species ($p < 0.001$, Table 1), except for $PM_{>10}$ at Site 1 ($p = 0.363$, Table 1). For all species, $PM_{>10}$ and $PM_{0.1-2.5}$ made up the greatest (66.4–89.2%) and smallest (4.4–10.0%) proportion of accumulated PM, respectively (Table 1).

The deposited $PM_{0.1-2.5}$, $PM_{2.5-10}$, $PM_{>10}$, and PM per unit green land were 8.3, 45.2, 18.7, and 16.1 $\mu g/cm^2$ ($PM_{0.1-2.5}$); 12.1, 131.0, 29.6, and 21.7 $\mu g/cm^2$ ($PM_{2.5-10}$); 170.4, 346.9, 249.1, and 209.7 $\mu g/cm^2$ ($PM_{>10}$), and 190.8, 523.1, 297.4, and 247.5 $\mu g/cm^2$ (PM), for *P. tomentosa*, *P. acerifolia*, *F. chinensis*, and *G. biloba*, respectively.

3.2. Differences in PM and Its Size Fractions among Sites

The PM deposited on leaves was significantly higher at Site 2 than that at Site 1 ($p < 0.001$, Table 1), with an increase of 40%, 180%, 40%, and 50% for *P. tomentosa*, *P. acerifolia*, *F. chinensis*, and *G. biloba*, respectively (Table 1). The corresponding increase at Site 2 compared with Site 1 was 20%, 260%, 30%, and 40% for $PM_{>10}$, 180%, 30%, 120%, and 110% for $PM_{2.5-10}$, and 140%, 80%, 120%, and 70% for $PM_{0.1-2.5}$. The ratio of $PM_{0.1-2.5}/PM_{0.1-10}$ was 0.38 at Site 1 and 0.36 at Site 2. Both are lower than that in ambient air, which was 0.82 at Site 1 and 0.69 at Site 2 during the growing season (May to October, 2014).

3.3. Morphological Structure of Leaf Surfaces

Table 2 and Figure 2 present the leaf surface structural properties of four studied tree species. *P. acerifolia* had the largest single-leaf area, followed by *P. tomentosa*, *F. chinensis*, and *G. biloba*. In terms of leaf wettability, all of the analyzed species, except the lower side of *G. biloba*, had a mean CA less than 90° (i.e., they were wettable) [22].

In the FESEM study, epicuticular wax was observed in tubular form (*G. biloba*, Figure 2j–l), or wax film (*P. tomentosa*, *P. acerifolia*, *F. chinensis*, Figure 2a–i). Wrinkled cuticles were observed on the lower side of *P. tomentosa* (Figure 2b,c), both surfaces of *P. acerifolia* (Figure 2d–f), lower side of *F. chinensis* (Figure 2h–i). The upper side of *F. chinensis* (Figure 2g) and *G. biloba* (Figure 2j). Stomata of the investigated species were either level with epidermal cells (*P. tomentosa*, Figure 2b,c), sunken (*F. chinensis*, Figure 2h,i, *G. biloba*, Figure 2k–l), or slightly elevated (*P. acerifolia*, Figure 2e,f). *P. tomentosa* had smaller stomata than the other species. However, the greatest stomatal aperture occurred on the foliage of *P. acerifolia*.

Table 1. Particulate matter and its size fractions accumulated on unit leaf area of trees ($\mu g/cm^2$), size fraction percentage (%, shown in parenthesis), and the amount ratios (%) as compared between Beijing Botanical Garden (Site 1) and Huangcun (Site 2).

Tree Species	Beijing Botanical Garden (Site 1) [a]				Huangcun (Site 2) [a]				Amount Ratios [b]			
	$PM_{0.1-2.5}$	$PM_{2.5-10}$	$PM_{>10}$	Total PM	$PM_{0.1-2.5}$	$PM_{2.5-10}$	$PM_{>10}$	Total PM	$PM_{0.1-2.5}$	$PM_{2.5-10}$	$PM_{>10}$	Total PM
P. tomentosa	3.9 ± 0.6 (4.4 ± 0.3)	5.7 ± 1.1 (6.4 ± 0.5)	80.0 ± 11.6 (89.2 ± 1.7)	89.6 ± 13.2	9.3 ± 0.8 (7.5 ± 0.8)	15.9 ± 1.7 (12.8 ± 0.9)	98.7 ± 4.7 (79.7 ± 1.4)	123.9 ± 5.5	2.4	2.8	1.2	1.4
P. acerifolia	14.2 ± 0.4 (8.6 ± 1.5)	41.2 ± 7.0 (25.0 ± 6.9)	109.1 ± 31.7 (66.4 ± 8.2)	164.5 ± 30.1	25.0 ± 4.9 (5.4 ± 1.0)	51.7 ± 7.6 (11.1 ± 1.7)	389.5 ± 15.2 (83.5 ± 2.6)	466.2 ± 26.9	1.8	1.3	3.6	2.8
F. chinensis	7.0 ± 2.0 (6.3 ± 0.7)	11.1 ± 2.5 (10.0 ± 0.9)	93.3 ± 23.2 (83.7 ± 2.2)	111.4 ± 27.4	15.6 ± 4.5 (10.0 ± 3.0)	24.1 ± 4.4 (15.4 ± 2.0)	117.3 ± 20.6 (74.6 ± 1.4)	157.0 ± 25.4	2.2	2.2	1.3	1.4
G. biloba	6.4 ± 2.3 (6.5 ± 2.4)	8.6 ± 1.2 (8.8 ± 1.3)	83.2 ± 2.7 (84.7 ± 1.5)	98.2 ± 1.8	10.8 ± 3.2 (7.3 ± 2.3)	18.0 ± 3.8 (12.1 ± 2.9)	119.5 ± 11.6 (80.6 ± 3.0)	148.3 ± 9.1	1.7	2.1	1.4	1.5

[a], mean ± SD; [b], mean values of PM and its size fractions deposited on leaves at Huangcun (Site 2) divided by those at Beijing Botanical Garden (Site 1).

Table 2. Means (±SD) of leaf surface structural properties in the Beijing Botanical Garden (Site 1) and Huangcun (Site 2).

Leaf Surface Structure Properties	Beijing Botanical Garden (Site 1)				Huangcun (Site 2)			
	P. tomentosa	*P. acerifolia*	*F. chinensis*	*G. biloba*	*P. tomentosa*	*P. acerifolia*	*F. chinensis*	*G. biloba*
Stomatal density ($/mm^2$)	217.1 ± 42.4	245.2 ± 56.8	205.0 ± 112.6	69.0 ± 22.4	278.8 ± 84.9	288.0 ± 26.3	423.5 ± 135.3	187.4 ± 37.8
Stomatal length (μm)	17.8 ± 1.5	29.1 ± 4.3	23.5 ± 4.6	23.1 ± 4.9	16.7 ± 3.0	20.9 ± 2.7	18.6 ± 2.7	15.2 ± 2.7
Stomatal width (μm)	7.4 ± 0.9	22.2 ± 3.6	11.3 ± 4.4	8.7 ± 2.2	5.3 ± 1.2	15.2 ± 1.5	8.0 ± 1.7	5.0 ± 1.6
Stomatal aperture (μm)	2.1 ± 1.0	7.3 ± 2.1	2.3 ± 1.2	n.d.	1.9 ± 0.7	6.3 ± 1.9	2.5 ± 0.6	n.d.
Stomatal with particle (%)	72.2 ± 31.5	66.4 ± 12.3	43.0 ± 25.9	14.6 ± 18.5	92.9 ± 12.6	89.9 ± 20.2	63.4 ± 18.0	11.0 ± 15.2
Contact angle (upper side) (°)	75.7 ± 4.6	67.4 ± 4.9	74.6 ± 6.2	74.0 ± 7.6	53.5 ± 7.0	63.8 ± 8.9	60.9 ± 3.1	68.8 ± 4.6
Contact angle (lower side) (°)	65.3 ± 3.2	67.1 ± 4.6	61.1 ± 7.8	103.1 ± 10.3	63.6 ± 8.1	56.3 ± 6.4	61.8 ± 5.7	93.2 ± 5.1
Single leaf area (cm^2)	84.9 ± 24.7	117.3 ± 22.3	24.8 ± 6.1	17.4 ± 3.6	53.6 ± 12.0	92.0 ± 18.8	19.4 ± 3.3	7.6 ± 2.9
Specific leaf area (cm^2/g)	89.6 ± 9.1	116.1 ± 9.6	119.0 ± 17.6	191.8 ± 7.3	106.8 ± 6.9	140.8 ± 17.1	129.7 ± 13.0	121.8 ± 10.8
Leaf length (cm)	10.1 ± 1.6	18.2 ± 2.3	9.8 ± 1.5	4.5 ± 0.6	7.8 ± 1.3	14.7 ± 1.3	8.6 ± 0.9	3.4 ± 0.6
Leaf width (cm)	9.3 ± 0.9	19.9 ± 1.0	4.6 ± 0.6	6.2 ± 1.0	7.1 ± 1.0	15.5 ± 1.7	3.9 ± 0.4	4.3 ± 1.2
Petiole length (cm)	8.6 ± 1.0	7.7 ± 1.1	1.6 ± 1.1	4.5 ± 1.7	5.2 ± 0.9	4.8 ± 1.2	1.2 ± 0.8	3.4 ± 1.1
Roughness (upper side)	1	3	3	4	2	3	3	4
Roughness (lower side)	3	4	4	5	3	4	4	5

n.d. indicates variables that were not found.

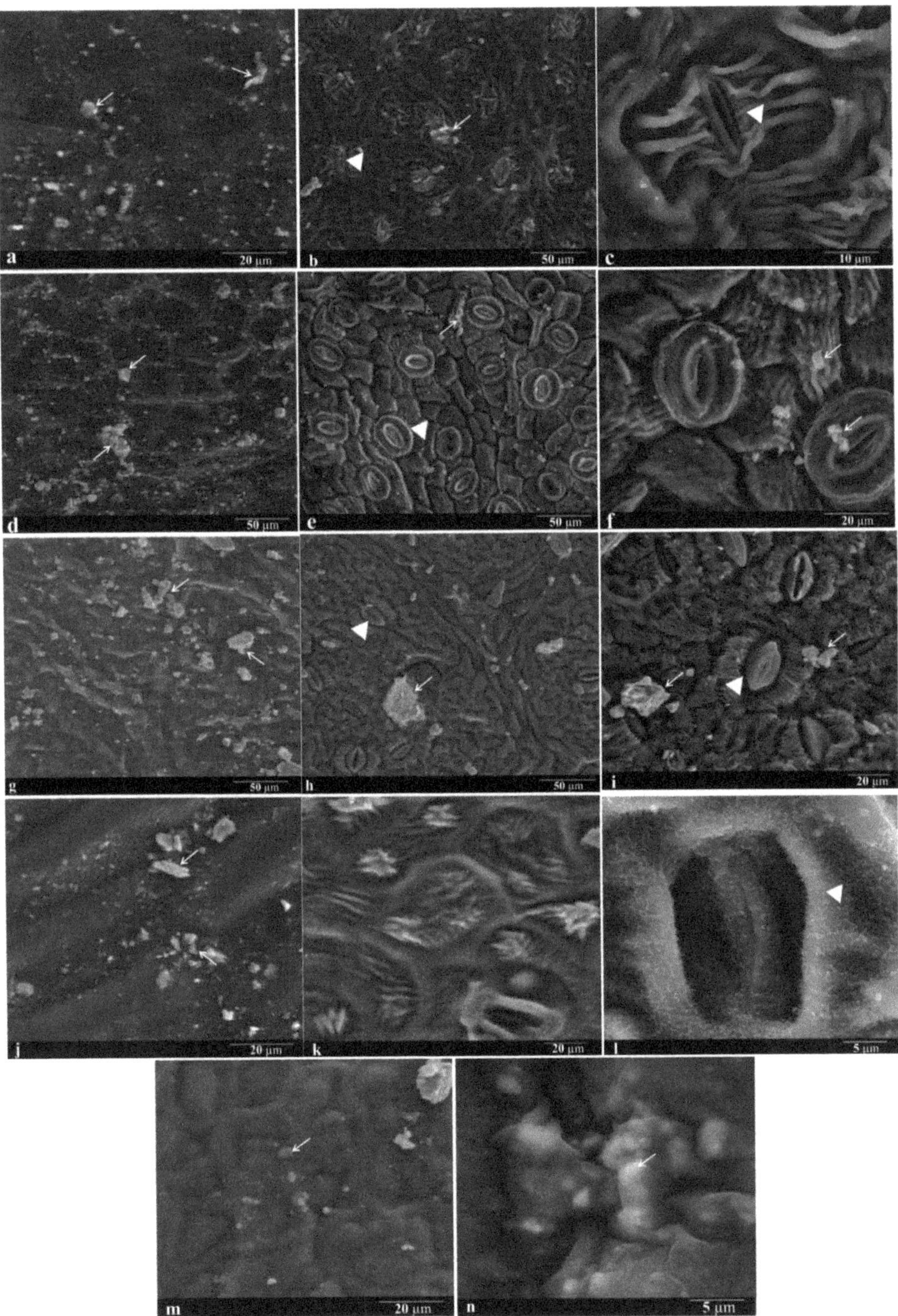

Figure 2. Scanning electron microscopy images of *Populus tomentosa* (**a**–**c**), (**d**–**f**), (**g**–**i**), and (**j**–**l**). **a**, **d**, **g**, **j**, **m**, and **n**, the upper side; **b**, **e**, **h**, and **k**, the lower side; **c**, **f**, **i**, and **l**, stomata. Particulate matter embedded in leaf epidermis (**m**: *Fraxinus chinensis*; **n**: *Ginkgo biloba*). Symbols indicate examples of stomata (triangle) and particulate matter (arrow).

3.4. Encapsulation of PM

PMs within cuticles were observed for *F. chinensis* (Figure 2m) and *G. biloba* (Figure 2n), but not for *P. tomentosa* and *P. acerifolia*. Encapsulation of PM within cuticles was not so common using FESEM observation. No visible damage was observed to either cuticle or epidermal cell. Wax encapsulated particles had diameters less than 6 μm, which was dominated by $PM_{2.5}$ (>90%).

3.5. The Effects of Leaf Structure on PM Accumulation

Stomatal aperture, stomatal width, leaf length, leaf width, stomatal density, and CA were crucial predictors for PM accumulation and its size fractions (Table 3, Figure 3). Among the 13 dependent variables, the sum of PC#1 and PC#2 was 71.8% (Figure 3). An increase in CA predicted a decrease in PM accumulation, while increased stomatal aperture, leaf length, leaf width, stomatal width, and stomatal density predicted an increase in PM accumulation (Table 3).

Table 3. Regression coefficient (B) and standardized regression coefficient (Beta) of the partial least squares regression model for the factors affecting leaf particulate matter capturing of four tree species.

	PM		$PM_{>10}$		$PM_{2.5-10}$		$PM_{0.1-2.5}$	
	B	Beta	B	Beta	B	Beta	B	Beta
Stomatal density	0.038	0.031	0.030	0.029	0.006	0.038	0.003	0.052
Stomatal length	0.057	0.002	0.015	0.001	0.122	0.034	0.018	0.012
Stomatal width	0.833	0.039	0.648	0.036	0.182	0.065	0.053	0.046
Stomatal aperture	2.757	0.060	2.254	0.058	0.470	0.077	0.162	0.064
Contact angle (upper side)	−0.473	−0.030	−0.362	−0.027	−0.074	−0.035	−0.041	−0.047
Contact angle (lower side)	−0.255	−0.035	−0.235	−0.039	−0.036	−0.037	−0.015	−0.039
Single leaf area	−0.036	−0.011	−0.039	−0.014	0.011	0.027	0.000	0.001
Specific leaf area	0.062	0.015	0.059	0.017	0.003	0.005	0.003	0.014
Leaf length	1.12	0.045	0.929	0.044	0.217	0.065	0.067	0.049
Leaf width	1.049	0.050	0.847	0.048	0.188	0.067	0.054	0.047
Petiole length	−0.194	−0.004	−0.138	−0.003	0.053	0.008	−0.040	−0.015
Roughness (upper side)	2.224	0.018	1.238	0.012	0.279	0.017	0.163	0.024
Roughness (lower side)	1.641	0.010	0.636	0.005	0.112	0.005	0.096	0.011

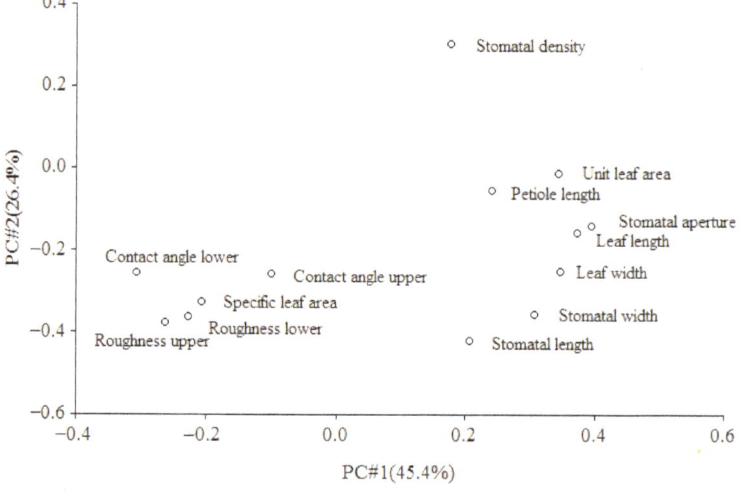

Figure 3. Loading plot of principal component analysis of the tree species with 13 dependent variables.

4. Discussion

4.1. Differences in PM Accumulation among Species

The plant species showed significant differences in PM accumulation among tree species and at different sites. In a study conducted by Sæbø et al. [11], they found that *Pinus mugo, P. sylvestris, Taxus media, Taxus baccata, Stephanandra incise,* and *B. pendula* showed higher PM accumulation. *Acer platanoides, Prunus avium,* and *Tilia cordata* showed lower PM accumulation. Zhang et al. [23] compared foliar PM retention and its size fractions of five plant species, *P. acerifolia* showed higher amount of PM accumulation ability than other species.

The differences in PM accumulation among species should be used for species selection during afforestation. However, tree size and LAI are important in determining the amount of PM accumulated per unit of land area, which is an index for PM retention ability and efficiency [24,25]. The relatively higher LAI of *P. acerifolia* further increased its PM removal potential, suggesting that this species could be used suitably as PM filters. The trees in Beijing could be damaged by air pollution; thus, the actual LAI was lower than healthy trees, which would influence the amount of accumulated PM. Among the investigated tree species, *P. acerifolia* is tolerant of air pollution, *G. biloba* and *P. tomentosa* are sensitive [26]. Therefore, we suggest that *P. acerifolia* should be used more in urban areas, especially in "hotspots" of PM pollution (e.g., the middle of the city, edges of streets/roads with heavy traffic loads). However, *G. biloba* and *P. tomentosa* should be installed in less polluted areas.

The contribution of PM in different size fractions to total PM on leaves decreased with the decreasing of PM diameter. Terzaghi et al. [27] found that large, coarse, and fine particles accounted for 87–95%, 5–12%, and 0.1–0.5% of the total PM for *Cornus mas, Acer pseudoplatanus,* and *Pinus pinea*. A study taken along a busy road in UK found the amount of PM captured on leaves increased with decreasing particle diameter [28]. These varied deposition level of PM with different aerodynamic diameter probably reflects the different aerodynamic properties of PM and their interactions with different leaf characteristics. Meanwhile, this difference can be ascribed to two groups of factors. The first group include PM concentration and compositions, plant species, and filtering materials (filter paper or membrane). The second group are mainly the used methods (measured mass or mass derived from particle number). Particles were assumed to be spherical and particle density equal for each size class [27].

4.2. Differences in PM Accumulation between Sites

More PM was found on the foliage of plants grown at heavily polluted site (Site 2) than less polluted site (Site 1). Deposition of pollutants depends on deposition velocity and pollutant concentration [29,30]. Micrometeorological conditions have been demonstrated to strongly impact on deposition dynamics, which may be partly explain the differences in PM accumulation between the two sites in our study. Furthermore, the deposition of PM on leaves depends on PM diameter. Deposition of PM with sizes of 0.1–1 μm is influenced by Brownian diffusivity and Stokes' law, and which is independent of size. For PM with diameter large than 1 μm, Stokes' law dominates the process of deposition, and which depends mostly on particle size. Larger particles will, therefore, deposit faster than smaller ones, either by sedimentation under the influence of gravity or by turbulent transfer resulting in impaction and interception [31], leading to a lower ratio of $PM_{0.1-2.5}/PM_{0.1-10}$ in Site 2 than the less polluted site.

In the present study, we observed a relatively higher percentage of fine (0.1–2.5 μm) and coarse (2.5–10 μm) size fractions in Site 2 than those in Site 1, while the large fractions were in lower percentages at Site 2. This finding seems to contrast with the results of Przybysz et al. [32], who obtained a relatively high rate of fine particles (6.2–9.8%) on plants at a rural site. Here, the lower ratio of $PM_{0.1-2.5}/PM_{0.1-10}$ accumulated on leaves, compared with that in ambient air, may be caused by three factors. First, some particles, elements, or ions are dissolved in water during the washing process [33]. Second, during the process of filtration, the membrane with pore sizes of 10 μm and (or) 2.5 μm may intercept some particles with diameters less than 10 μm and (or) 2.5 μm after saturation [34]. Third, the leaf cuticle encapsulated some particles, especially tiny ones [27].

4.3. Importance of Leaf Structure for PM Accumulation

Urban trees accumulating PM is a complex and poorly characterized process which is influenced by many factors such as leaf surface micro-, and macro-structure (e.g., leaf wettability, stomatal density, leaf area, leaf roughness, and the shape and amount of trichomes) [11,23,24,35]. Research has shown that leaves with small size, low CAs, low stomatal density, high stomatal conductance, and higher amount of leaf hairiness can capture more PM [11,18,19,21,31]. In this study, we also found that high CAs decreased the amount of captured PM. The contact area between a particle and the underlying leaf surface is considerably reduced on surfaces with high CAs. Consequently, the physical adhesion forces between particles and leaf surfaces are reduced, leading to a lower PM accumulation [19]. However, we also found that increased stomatal aperture, density, and width increased the amounts of captured PM. Large stomatal density, aperture, and width could result in increased transpiration which can make particles more deliquescent. And as a consequence, deposition rates increase [36]. Transpiration of water through stomata can cool leaf surfaces, but increase PM deposition by thermophoresis [18], which may also partly explain the higher ability of leaves to capture PM under more polluted conditions than conparatively less polluted areas.

We found that *P. acerifolia*, the species with the largest leaves, had the highest PM accumulation among the investigated species. This is in opposition to the findings of previous studies, in which the authors found small leaves increased the number of captured particles [18,28]. According to Nobel [37], particles in the air could more easily collide with small leaves than large and flat leaves, which have thicker boundary layers. This contrasting finding may be caused by the microstructure of *P. acerifolia* leaves, which had a rough surface that can influence the boundary layer [38].

Some PM with diameters less than 6 μm (mainly $PM_{2.5}$) were encapsulated in cuticles of leaves of *F. chinensis* and *G. biloba*, but not for *P. acerifolia* and *P. tomentosa*, by FESEM observation. These results suggest that the potential of PM embedded in wax layer depends on the quantity of wax as well as the composition and structure of the epicuticular wax layer; these are species-specific characteristics [39]. Terzaghi et al. [27] found that particles with diameters less than 10.6 μm were encapsulated into cuticles. The amount of encapsulation and the capacity of leaves to capture PM changed over time. They attributed this to the degradation of cuticular waxes, from a perfect wax crystal to an amorphous one. Dzierżanowki et al. [39] demonstrated that large particles appeared mainly on the leaf surface rather than in the wax layer of some plant species. The encapsulated particles always had small diameters and could not be easily washed off during rain events or dislodged by wind. If most of the accumulated PM is immobilized which can be considered to be beneficial in the planning of PM phytoremediation. However, if the washed- or blown-off PM is considered to be filter cleaning, leaving the leaves ready for additional deposition. This process may result in underestimating the PM removal effect [23]. The dynamics of deposition, including the amounts of PM washed-off by rain and blown-off by wind, need further investigation.

5. Conclusions

(i) The amounts of accumulated PM differed significantly among species, in the order of *P. acerifolia* > *F. chinensis* > *G. biloba* > *P. tomentosa*. Most of the accumulated PM belonged to the largest fraction (>10 μm). Some PM was encapsulated in cuticles of *F. chinensis* and *G. biloba*, and which was dominated by $PM_{2.5}$ (>90%).

(ii) Trees at polluted site had higher rates of PM accumulation and higher percentage of fine and coarse fractions than less polluted site. With the increase of pollution level, the PM retention ability of tree species increased with the decrease of particle size, indicating that plant leaves could accumulate fine particles and purify local air.

(iii)   Leaf structures affect PM accumulation and its size fractions. Large leaves, along with low stomatal aperture, width, and density, as well as low CA, all resulted in increased PM capture.

Author Contributions: Conceptualization, H.W., H.S. and Y.W.; methodology, H.W. and Y.W.; validation, H.W.; investigation, H.W., Y.X., J.Y. and B.X.; resources, H.W. and Y.X.; data curation, H.W. and Y.X.; writing—original draft preparation, H.W., J.Y. and B.X.; writing, review and editing, H.W., H.S. and Y.W.; visualization, H.W., J.Y. and B.X.; supervision, H.S. and Y.W.; project administration, H.W.; funding acquisition, H.W. All authors have read and agreed to the published version of the manuscript.

Funding: This research was funded by the Scientific Research Program of the Education Department of Shaanxi Provincial Government, grant number 20JS082.

Acknowledgments: We are grateful to Wenyan Yang, Tsinghua University, for assistance with the field emission scanning electron microscopy. We also are grateful to the Key Lab of Northwest Water Resources and Environment Ecology of Ministry of Education for providing the goniometer.

Conflicts of Interest: The authors declare no conflict of interest.

References

1.   Beijing Municipal Ecological and Environmental Monitoring Center Ambient Air Quality Daily Report. Available online: http://www.bjmemc.com.cn/ (accessed on 17 May 2021). (In Chinese)

2.   Liu, J.; Yin, H.; Tang, X.; Zhu, T.; Zhang, Q.; Liu, Z.; Tang, X.; Yi, H. Transition in air pollution, disease burden and health cost in China: A comparative study of long-term and short-term exposure. *Environ. Pollut.* **2021**, *277*, 116770. [CrossRef] [PubMed]

3.   Mukherjee, A.; Agrawal, M. World air particulate matter: Sources, distribution and health effects. *Environ. Chem. Lett.* **2017**, *15*, 283–309. [CrossRef]

4.   Dockery, D.W.; Pope, C.A. Acute Respiratory Effects of Particulate Air Pollution. *Annu. Rev. Public Health* **1994**, *15*, 107–132. [CrossRef]

5.   Tchepel, O.; Dias, D. Quantification of health benefits related with reduction of atmospheric PM10levels: Implementation of population mobility approach. *Int. J. Environ. Heal. Res.* **2011**, *21*, 189–200. [CrossRef]

6.   Wang, H.X.; Maher, B.A.; Ahmed, I.A.M.; Davison, B. Efficient Removal of Ultrafine Particles from Diesel Exhaust by Selected Tree Species: Implications for Roadside Planting for Improving the Quality of Urban Air. *Environ. Sci. Technol.* **2020**, *53*, 6906–6916. [CrossRef]

7.   Platel, A.; Privat, K.; Talahari, S.; Delobel, A.; Dourdin, G.; Gateau, E.; Simar, S.; Saleh, Y.; Sotty, J.; Antherieu, S.; et al. Study of in vitro and in vivo genotoxic effects of air pollution fine ($PM_{2.5-0.18}$) and quasi-ultrafine ($PM_{0.18}$) particles on lung models. *Sci. Total Environ.* **2020**, *711*, 134666. [CrossRef]

8.   Ministry of Environmental Protection of the People's Republic of China. *Ambient Air Quality Standards (GB 3095-2012)*; China Environmental Science Press: Beijing, China, 2012; p. 3. (In Chinese)

9.   Zhang, M.; Han, L.H.; Liu, B.X.; Wang, Q.; Cao, Y. Evolution of $PM_{2.5}$ and its components during heavy pollution episodes in winter in Beijing. *China Environ. Sci.* **2020**, *40*, 2829–2838. (In Chinese with English Abstract)

10.   Zhang, X.Y.; Lyu, J.Y.; Han, Y.J.; Sun, N.X.; Sun, W.; Li, J.M.; Liu, C.J.; Yin, S. Effects of the leaf functional traits of coniferous and broadleaved trees in subtropical monsoon regions on $PM_{2.5}$ dry deposition velocities. *Environ. Pollut.* **2020**, *265*, 114845. [CrossRef]

11.   Sæbø, A.; Popek, R.; Nawrot, B.; Hanslin, H.M.; Gawronska, H.; Gawronski, S.W. Plant species differences in particulate matter accumulation on leaf surfaces. *Sci. Total Environ.* **2012**, *427–428*, 347–354. [CrossRef]

12.   Fowler, D.; Cape, J.N.; Unsworth, M.H.; Mayer, H.; Crower, J.M.; Jarvis, P.G.; Gardiner, B.; Shuttleworth, W.J. Deposition of atmospheric pollutants on forests. *Philos. Trans. R. Soc. B Biol. Sci.* **1989**, *324*, 247–265. [CrossRef]

13.   McDonald, A.G.; Bealey, W.J.; Fowler, D.; Dragosits, U.; Skiba, U.; Smith, R.I.; Donovan, R.G.; Brett, H.E.; Hewitt, C.N.; Nemitz, E. Quantifying the effect of urban tree planting on concentrations and depositions of PM_{10} in two UK conurbation. *Atmos. Environ.* **2007**, *41*, 8455–8467. [CrossRef]

14.   Nowak, D.J.; Hirabayashi, S.; Ellis, E.; Greenfield, E.J. Tree and forest effects on air quality and human health in the United States. *Environ. Pollut.* **2014**, *193*, 119–129. [CrossRef] [PubMed]

15.   Wang, X.S.; Teng, M.J.; Huang, C.B.; Zhou, Z.X.; Chen, X.P.; Xiang, Y. Canopy density effects on particulate matter attenuation coefficients in street canyons during summer in the Wuhan metropolitan area. *Atmos. Environ.* **2020**, *240*, 117739. [CrossRef]

16.   Uni, D.; Katra, I. Airborne dust absorption by semi-arid forests reduces PM pollution in nearby urban environments. *Sci. Total Environ.* **2017**, *598*, 984–992. [CrossRef] [PubMed]

17.   Escobedo, F.J.; Nowak, D.J. Spatial heterogeneity and air pollution removal by an urban forest. *Landsc. Urban Plan* **2009**, *90*, 102–110. [CrossRef]

18. Räsänen, J.V.; Holopainen, T.; Joutsensaari, J.; Ndam, C.; Pasanen, P.; Rinnan, Å.; Kivimäenpää, M. Effects of species-specific leaf characteristics and reduced water availability on fine particle capture efficiency of trees. *Environ. Pollut.* **2013**, *183*, 64–70. [CrossRef] [PubMed]

19. Wang, H.X.; Shi, H.; Li, Y.Y.; Yu, Y.; Zhang, J. Seasonal variations in leaf capturing of particulate matter, surface wettability and micromorphology in urban tree species. *Front. Environ. Sci. Eng.* **2013**, *7*, 579–588. [CrossRef]

20. Yue, C.; Cui, K.D.; Duan, J.; Wu, X.Y.; Yan, P.B.; Rodriguez, C.; Fu, H.M.; Deng, T.; Zhang, S.W.; Liu, J.Q.; et al. The retention characteristics for water-soluble and water-insoluble particulate matter of five tree species along an air pollution gradient in Beijing, China. *Sci. Total Environ.* **2021**, *767*, 145497. [CrossRef] [PubMed]

21. Zhang, W.K.; Zhang, Z.; Meng, H.; Zhang, T. How does leaf surface micromorphology of different trees impact their ability to capture particulate matter? *Forests* **2018**, *9*, 681. [CrossRef]

22. Aryal, B.; Neuner, G. Leaf wettability decreases along an extreme altitudinal gradient. *Oecologia* **2010**, *162*, 1–9. [CrossRef]

23. Zhang, L.; Zhang, Z.Q.; Chen, L.X.; McNulty, S. An investigation on the leaf accumulation-removal efficiency of atmospheric particulate matter for five urban plant species under different rainfall regimes. *Atmos. Environ.* **2019**, *208*, 123–132. [CrossRef]

24. Wang, H.X.; Shi, H. Particle retention capacity, efficiency, and mechanism of selected plant species: Implications for urban planting for improving urban air quality. *Plants* **2021**, *10*, 2109. [CrossRef] [PubMed]

25. Weerakkody, U.; Dover, J.W.; Mitchell, P.; Reiling, K. Quantification of the traffic-generated particulate matter capture by plant species in a living wall and evaluation of the important leaf characteristics. *Sci. Total Environ.* **2018**, *635*, 1012–1024. [CrossRef] [PubMed]

26. Wang, H.X.; Wang, Y.H.; Shi, H. *Abatement of Air Particulate Pollutants by Urban Greening Plants*, 1st ed.; China Forestry Publishing House: Beijing, China, 2020; pp. 90–91.

27. Terzaghi, E.; Wild, E.; Zacchello, G.; Cerabolini, B.E.L.; Jones, K.C.; Guardo, A.D. Forest filter effect: Role of leaves in capturing/releasing air particulate matter and its associated PAHs. *Atmos. Environ.* **2013**, *74*, 378–384. [CrossRef]

28. Weerakkody, U.; Dover, J.W.; Mitchell, P.; Reiling, K. Topographical structures in planting design of living walls affect their ability to immobilise traffic-based particulate matter. *Sci. Total Environ.* **2019**, *660*, 644–649. [CrossRef] [PubMed]

29. Nowak, D.J.; Hirabayashi, S.; Bodine, A.; Hoehn, R. Modeled $PM_{2.5}$ removal by trees in ten U.S. cities and associated health effects. *Environ. Pollut.* **2013**, *178*, 395–402. [CrossRef] [PubMed]

30. Tallis, M.; Taylor, G.; Sinnett, D.; Freer-Smith, P. Estimating the removal of atmospheric particulate pollution by the urban tree canopy of London, under current and future environments. *Landsc. Urban Plan* **2011**, *103*, 129–138. [CrossRef]

31. Freer-Smith, P.H.; Beckett, K.P.; Taylor, G. Deposition velocities to *Sorbus aria*, *Acer campestre*, *Populus deltoids × trichocarpa* 'Beaupré', *Pinus nigra* and × *Cupressocyparis leylandii* for coarse, fine and ultra-fine particles in the urban environment. *Environ. Pollut.* **2005**, *133*, 157–167. [CrossRef]

32. Przybysz, A.; Sæbø, A.; Hanslin, H.M.; Gawroński, S.W. Accumulation of particulate matter and trace elements on vegetation as affected by pollution level, rainfall and the passage of time. *Sci. Total Environ.* **2014**, *481*, 360–369. [CrossRef]

33. Wu, F.Y.; Wang, W.; Man, Y.B.; Chan, C.Y.; Liu, W.X.; Tao, S.; Wong, M.H. Levels of $PM_{2.5}/PM_{10}$ and associated metal(loid)s in rural households of Henan Province, China. *Sci. Total Environ.* **2015**, *512–513*, 194–200. [CrossRef]

34. Yang, J.; Wang, H.X.; Xie, B.Z.; Wang, Y.H.; Shi, H. Quantity and mass of particulate matter retained on tree leaves and determination methods. *J. Ecol. Rural Environ.* **2015**, *31*, 440–444, (In Chinese with English Abstract).

35. Kardel, F.; Wuyts, K.; Maher, B.A.; Hansard, R.; Samson, R. Leaf saturation isothermal remanent magnetization (SIRM) as a proxy for particulate matter monitoring: Inter-species differences and in-season variation. *Atmos. Environ.* **2011**, *45*, 5164–5171. [CrossRef]

36. Burkhardt, J.; Koch, K.; Kaiser, H. Deliquescence of deposited atmospheric particles on leaf surfaces. *Water Air Soil Poll.* **2001**, *1*, 313–321. [CrossRef]

37. Nobel, P.S. *Physicochemical and Environmental Plant Physiology*, 4th ed.; Academic Press: New York, NY, USA, 2009; pp. 228–275. [CrossRef]

38. Barber, J.L.; Thomas, G.O.; Kerstiens, G.; Jones, K.C. Air-side and plant-side resistances influence the uptake of airborne PCBs by evergreen plants. *Environ. Sci. Technol.* **2002**, *36*, 3224–3229. [CrossRef]

39. Dzierżanowski, K.; Popek, R.; Gawrońska, H.; Sæbø, A.; Gawroński, S.W. Deposition of particulate matter of different size fractions on leaf surfaces and in waxes of urban forest species. *Int. J. Phytoremediat.* **2011**, *13*, 1037–1046. [CrossRef]

forests

MDPI

Article

Analysis of the Noise Pollution in the Bielański Forest NATURA 2000 Area in Light of Existing Avifauna (Warsaw, Poland)

Agata Pawłat-Zawrzykraj *, Paweł Oglęcki and Konrad Podawca

Institute of Environmental Engineering, Faculty of Civil and Environmental Engineering, Warsaw University of Life Science—WULS, ul. Nowoursynowska 166, 02-786 Warsaw, Poland; pawel_oglecki@sggw.edu.pl (P.O.); konrad_podawca@sggw.edu.pl (K.P.)

* Correspondence: agata_pawlat_zawrzykraj@sggw.edu.pl

Abstract: There is no doubt that the NATURA 2000 network has been one of the most relevant tools for nature protection. However, the designated areas within the borders of large cities are subjected to many threats. Traffic noise takes on a very important role in this subject, posing the question of whether NATURA 2000 areas should be located in urban areas strongly affected by noise pollution. This particular topic was exposed and analysed at Bielański Forest NATURA 2000 (PLH140041), located in northern Warsaw, where changes in noise distribution for the years 2007–2017 were examined and described by two types of indicators (Lden and Lnight). The data sources used for the analysis were city road noise maps for 2007, 2012 and 2017. Additionally, sound intensity measurements were taken in two separate groups of hot-spots. The first of these comprised locations determined based on an inventory of avifauna; they represented different habitat types, and were characterized by the highest bird activity. The second group of hot-spots consisted of those designated along roads in order to identify the main sources of traffic noise. The obtained results confirmed the high noise-absorbing ability of the existing vegetation. The avifauna surveys covered 19 forest bird species. Five of them were considered to be the most valuable and rare elements of the local avifauna, whereas 14 appeared to be key species for the functioning of biocenosis and, at the same time, determine the uniqueness of the ecosystem. The study showed that the type of habitat rather than differences in noise levels determines the distribution and abundance of key species. Therefore, there is a necessity to focus on actions that guarantee the maintenance of the existing status in order to counteract habitat deterioration. The investigation confirmed the feasibility of creating these kinds of natural protected areas in large cities, despite their exposure to noise pollution.

Keywords: urban forest; NATURA 2000 area; soundscape; birds; biological diversity

Citation: Pawłat-Zawrzykraj, A.; Oglęcki, P.; Podawca, K. Analysis of the Noise Pollution in the Bielański Forest NATURA 2000 Area in Light of Existing Avifauna (Warsaw, Poland). *Forests* **2021**, *12*, 1316. https://doi.org/10.3390/f12101316

Academic Editors: Manuel Esperon-Rodriguez and Tina Harrison

Received: 20 July 2021
Accepted: 22 September 2021
Published: 26 September 2021

1. Introduction

Urban NATURA 2000 sites and other areas with unique natural values under legal protection are important parts of the green area systems of cities [1]. They play a key role in the shaping of urban biodiversity, as well as climate and hydrological conditions [2,3]. However, the location of NATURA 2000 areas in the city structure makes them vulnerable to many negative factors—both internal and external. The main internal (direct) threats that are also relevant to Warsaw NATURA 2000 sites [4,5] include: inadequate maintenance, the destruction of plants, the introduction of alien and invasive species that supplant native ones, and damage to and the killing or trading of protected wild fauna and flora species. These threats are often related to the additional function of these areas—recreation. Water, soil, air, light and noise pollution are the main indirect factors negatively impacting the environment. Other factors exhibiting the same effect are associated with land use changes, especially the growing urbanization of the areas adjacent to high value nature sites. New developments and a decrease in the biologically active surface have an adverse effect on water conditions, e.g., a decrease of water retention and a lowering of the ground water levels. The main consequence of these common phenomena is the deterioration of the living conditions and natural habitats of some biota (especially birds).

The study focuses on one of the aforementioned threats—noise pollution. The evaluation of urban diversification in regards to noise pollution is crucial, not only for noise mitigation but also for the assessment of acoustic comfort, which is integral to the overall environmental quality [6]. The issue can be discussed on different scales, including agglomeration, the entire city, individual green areas, or even a single tree. On the urban and local level, there is a strong correlation between noise pollution and land cover. However, greener cities may not always be quieter. This is also a matter of the ratio of green space and built-up surfaces, as well as certain characteristics of green areas. It has been proven that lower noise levels may be achieved in cities with a higher extent of porosity and green space coverage [7]. The higher the density of the greenery that absorbs the acoustic waves expressed by the total surface of leaves the greater the noise level reduction [8]. This is particularly apparent in areas with a high share of forests and agricultural areas in combination with a small share of road network infrastructure [7]. At the same time, grass-covered grounds (lawns) appear to be rather inefficient in terms of noise reduction [8,9]. According to Van Renterghem [10], vegetation can reduce the noise level even by as much as 10 dB.

In cities, the state of the noise pollution from road traffic is presented on noise maps. They are the main source of data for the evaluation of the spatial extent and intensity of this phenomenon, and they provide us with valuable input into the decision making process in urban land use planning and management. One of the questions is what noise level can be acceptable for city green areas, such as parks and forests.

According to the World Health Organization [11], the noise level in recreational areas should not exceed 55 dB. The permissible levels of noise specified in the Polish Act concerning the average maximum traffic noise levels for recreational areas depend on the time of day and are limited to 68 dB during the day and 59 dB at night [12]. The issue is mostly discussed in relation to the human annoyance response, which is, however, a matter of subjective perception, depending on personal preference and opinions [13], specific types of sounds [14,15], the presence of trees natural features, and the tranquility of the area. Therefore, the results of research regarding the respondents' perception of the soundscape of green areas may differ slightly from the results obtained from noise maps and the measurements of sound pressure levels [16].

It can be assumed that noise pollution affects not only humans but also other users of urban spaces, such as wild animals, especially birds. One of the most important elements of the natural environment—from both the human perspective and that of ecosystem ecology—are undoubtedly birds. More than a hundred species (not counting incidental migrants) of these vertebrates have been observed in the cities of Central Europe [17]. It is not a coincidence that one of the two fundamental directives enabling the creation of NATURA 2000 areas is called the Bird Directive; birds are a group of animals of great interest. Their importance for the overall condition of ecosystems (and urban ecosystems) and the indicator of their general well-being cannot be overestimated [18]. The condition of urban avifauna can vary greatly depending on the geographic location, proximity to natural enclaves, availability of free migration corridors and many other factors, both biotic and abiotic—e.g., population size and the intensity of green infrastructure [19]. However, the vast majority of researchers point out that the general trend for urban avifauna is a decline in the richness of species and, consequently, biological diversity [17,20–22]. Again, various factors are cited as the cause, among which are the increasing noise levels [23–25]. Although the effect of noise on individual species varies, it has a negative effect on most species, even if some have managed to adapt to it. Birds respond to increased noise by becoming more skittish and stressed, and may be less likely to feed and breed. Noise may also hinder their ability to perceive danger. Regular exposure to urban noise may cause changes in bird communities and influences local distribution patterns [26]. The effects of noise are of great importance in urban areas because they can lead to a global homogenization of bird communities, particularly if certain species fail to occupy noisy areas [27]. The increase in noise may also lead to urban bird communities being dominated by a limited number of noise-tolerant species [28]. Quoting Manzanares and Lucia [28], one motivation to study the effect of noise on urban bird communities is to promote urban avian diversity. Thus, it is necessary to understand how birds respond to anthropogenic noise. In the literature, different approaches are being used to achieve this. One of them consists of studying the relation between anthropogenic noise and the structure of birdsong which is important in terms of territorial defense and female attraction [24,26–29], Taking into account the above facts, special attention should be paid to green areas within cities in terms of their importance as refuges for avifauna.

The main objective of the study is to analyse changes in traffic noise pollution in Bielański Forest, especially in its most valuable part—the NATURA 2000 site, in relation to the diversity of existing avifauna. The following questions were formulated: (1) What are the temporal and spatial variability of the traffic noise pollution in the area of Bielański Forest? (2) What is the distribution of the bird species composition, with a particular focus on rare species in relation to noise pollution? (3) What other factors may be associated with the current state of the bird communities in Bielański Forest?

2. The Study Area

The study concerns an area called Bielański Forest (the general area of the study), located in the northern part of Warsaw, in the immediate vicinity of the Vistula river. Bielański Forest is a remainder of the old Mazovian Primeval Forest, which covers almost 98% of the area of study. Two protected nature sites which overlap partly are designated within the forest: the Nature Reserve—Bielański Forest, with a total area of 130.35 ha and the NATURA 2000 (PLH140041) Special Area of Conservation, with an area of 129.84 ha [30]. The present research covered the latter.

Bielański Forest is an important part of the Warsaw Natural System (WNS). This is a concept that was introduced in Warsaw's Spatial Policy [31] in order to stabilize and enhance environmental processes in the city. The system consists of core and supporting areas, which are the major and most valuable structural elements for the climatic, hydrological and biological performance of the city. It is worth mentioning that core areas consist of areas with the highest natural and landscape values, being part of supra-regional natural connections, such as: the flood terraces of the Vistula River with characteristic plant communities, large and compact forest complexes, valuable meadows and grasslands, the northern and southern section of the Warsaw Escarpment, the most valuable areas of urban greenery, and urbanized areas with a biologically active surface area of no less than 70%. Supporting areas consist of areas of high natural and landscape values, supplementing the structure of the core areas and maintaining their spatial and functional connectivity, such as selected areas of urban greenery and forests, as well as selected urbanized areas with a biologically active surface area of no less than 60% [31]. The main elements of the system are linked by areas that form basic ecological connections. Bielański Forest is one of the core areas of WNS (Figure 1B). The importance of this area in the system and its level of biological diversity are dependent not only on the existing natural environmental resources but also on spatial connections with other biologically active terrains within the city, and those located in its vicinity. Unfortunately, the area is cut off from the Vistula valley (a supra-regional ecological corridor, Special Protection Area of Birds-SPAs NATURA 2000—PLB140004) by Wybrzeże Gdyńskie Street—one of the main roads in this part of the city (Figure 1).

Forests cover about 17% of the total city area [32]. The distribution of forests is uneven. They are located mainly in the urban periphery. Bielański Forest is an important link in the natural system of the western outskirts of Warsaw. In its direct neighborhood there is a vast forest complex called Młociński Forest, and about 15 km further is the buffer zone of Kampinoski National Park. Animals which are intentionally migrating or straying from both of these areas, e.g., moose, can be found in Bielański Forest.

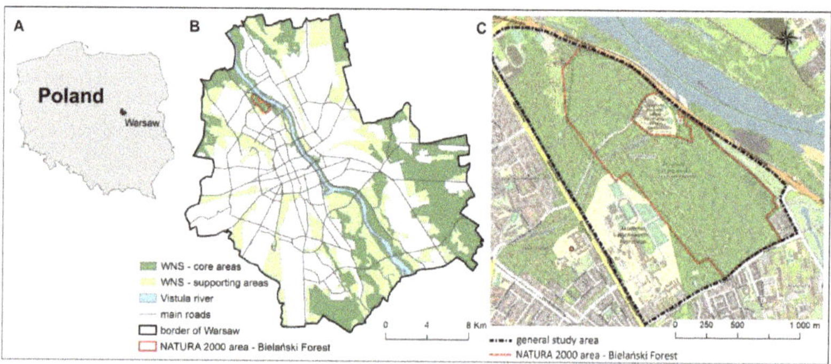

Figure 1. Area of the study (own elaboration): (**A**)—Warsaw on the map of Poland, (**B**)—Bielański Forest as an element of Warsaw Natural System (WNS), (**C**)—map of Bielański Forest.

The study area is diverse in terms of its geomorphology, water conditions and the structure of its habitats. There are four terraces running parallel to the Vistula River: a flood terrace, a higher terrace (an alluvial one), an ice valley of the Vistula, and a moraine plateau called Bielański terrace. The area of the flood terrace is closest to the Vistula, covering the valley of Rudawka Stream. The groundwater level in this area depends on the water level of the Vistula and ranges from 0.0 to 2.0 m below the ground level [5,33]. The area is covered with an elm-ash forest (Ficario-Ulmetum), which is rarely found in urban areas. The current water conditions are within the tolerance levels of the plant species occurring in this area. It is the most natural part of Bielański Forest. The area of the higher flood terrace is mostly covered by the oak-hornbeam forest (fertile variant, Tilio-Carpinetum) and, in a few locations, by an elm-ash forest (a drier variant of Ficario-Ulmetum with hornbeam). Numerous old trees can be found in these parts. In the southern part of the site, above the flood terrace, there is a sandy terrace of the Vistula proglacial valley. In some places, small elongated hills, now overgrown with pine trees, are still visible. The moraine plateau is the highest part of the site in terms of elevation. The terrace is bordered by an escarpment rising 6–15 m above the Vistula River valley. The groundwater level lies at a depth of 5–10 p.p.t., and locally even deeper [33]. This part of the area is covered with high oak-hornbeam forest of a drier variant (Tilio-Carpinetum Typicum). Some of the old oak stands are about 300–400 years old.

The water conditions in the studied area have been transformed due anthropogenic pressure, which is associated mainly with the development of multi-family houses and service establishments, as well as the construction and modernization of road and technical infrastructure in the surroundings of the study area. An irreversible decrease of the groundwater level had already been indicated in a study carried out in 1994 [4]. It concerned mainly the Bielański terrace (about a 1.5 m ground water level drop) and, to a lesser degree, the higher flood terrace (about 0.5 m). The smallest changes were observed in the area of the flood terrace. The threats of a disturbance in the hydrological conditions and falling trees leading to changes in the structure and composition of habitat's species as well as a loss of animal habitats, were pointed out by Skorupski et al. [4]. Therefore, the area was assessed as being susceptible to degradation [4]. Changes in the structure of plant communities mainly involving degradation towards drier variants of the oak-hornbeam forest indicate that this negative trend is continuing. Nevertheless, according to the natural evaluation of the landscape, the study area obtained a very high landscape value category [34].

It is estimated that Bielański Forest is a habitat for about 1000 species of invertebrates (the most valuable of which are great the capricorn beetle—*Cerambyx cerdo* and the hermit beetle—*Osmoderma eremita*), and 20 species of mammals—squirrels, several species of bats, deer, foxes, wild boars, and occasionally moose The least numerous are reptiles and amphibians, which have suffered from the progressive drainage of the area. Birds are the largest group of local fauna [4].

3. Materials and Methods

The study concerns the following two issues:

1. Changes in the distribution of noise pollution in the area of Bielański Forest with a particular consideration of the NATURA 2000 site.
2. The state of the avifauna in the most valuable part of the area—the NATURA 2000 site.

3.1. The Noise Pollution Analysis

The legal basis for the analysis of changes in noise pollution regarding land-use planning and management is the Ordinance of the Minister of the Environment of 14 June 2007 on the permissible noise levels in the environment [12]. It refers to the environmental noise to which primarily humans are exposed, in particular noise sensitive buildings and areas, such as residential areas, schools, hospitals, public parks and recreational and leisure areas. Despite the fact that the study area is under nature conservation as a nature reserve and a NATURA 2000 area, it is also a popular place of recreation and rest for the citizens of Warsaw. The impact of urban pressure on such areas in Warsaw and the Warsaw Functional Area is high [5,35,36]. In order to establish a long-term and sustainable spatial policy concerning the maintenance of existing recreational and leisure areas and the designation of new ones, the Lden and Lnight indicators are used. The first of these refers to the long-term average sound level calculated for all of the days of the year during daytime-evening-night (Lden), whereas the latter is calculated for night time. The current maximum long-term noise levels for recreational and leisure areas affected by roads and railways are 68 dB for Lden and 59 for Lnight, while in the case of noise affected by other facilities, they are 55 dB for Lden and 45 for Lnight [12].

In order to characterize the changes in the noise pollution for the analyzed area during the period of 2012–2017, surface and percentage changes in the areas exposed to different sound levels were determined. Similar analyses were performed by the authors regarding residential areas for road and railway noise [37–39]. The main sources of noise pollution in the area of the study are two roads located along the eastern and southern borders (Wybrzeże Gdyńskie Street and Podleśna Street).

The original data were gained from road noise maps of Warsaw for 2007, 2012 and 2017 [40]. The noise maps for these years were made on the basis of the "French" NMPB-Routes-96 method (SETRA-CERTU-LCPC-CSTB) and the French XPS 31–133 standard, in accordance with Directive 2002/49/EC and the methodological foundations of geographic information science [41,42]. The maps were developed using numerical modeling based on point measurements carried out at the altitude of 4 m above ground level and for a calculation grid with vertical steps of 15 m [43].

The spatial data analysis for Bielański Forest was carried out using ArcGIS version 10.7.1, including the following procedure:

- generating rasters with road noise maps for 2007, 2012 and 2017 for Lden and Lnight indicators;
- the georeferencing of the rasters according to the PUWG_92 mapped coordinate system;
- vectorization—changing raster graphics into vector graphics;
- the determination of changes in the alignment of the selected noise bands in 2007, 2012 and 2017, using the data management tool;
- the designation of areas of changes regarding the assumed noise levels using the coefficient for area variation, according to Formula (1).

$$W_{CA17-12(12-07)} = \left[\frac{A_{Lden\ (Lnight)2017(2012)} - A_{Lden(Lnight)2012(2007)}}{A_{a(n)}} \right] \times 100(\%) \tag{1}$$

where $A_{Lden(Lnight)2017(2012)}$ is the area affected by noise pollution at sequential ranges of values according to the Lden or Lnight indicators in 2017 or 2012 (ha), $A_{Lden(Lnight)2012(2007)}$ is the area affected by noise pollution at sequential ranges of values according to Lden or Lnight indicators in 2012 or 2007 (ha), A_a is the basic area of study (ha), and A_n is the NATURA 2000 area (ha).

A negative value of the W_{CA} index means a decrease in the area within the analyzed range of nuisance, and a positive value indicates an increase. In the case of the W_{CA} index for an Lnight noise level up to 60 dB, positive values are considered favorable while negative values stand for a negative trend. In the case of the index calculated for Lden, on the other hand, positive values for a noise level above 60 dB indicate a deterioration of conditions, and negative values indicate an improvement of the acoustic conditions.

In order to supplement the information on noise pollution which is available from the noise maps, field measurements of the sound levels were taken. This was accomplished by utilizing the digital sound analyzer model DSA-50 which allows the following characteristics to be measured:

- $L_{AI\,min}$—the minimum value of the effective sound level,
- $L_{AI\,max}$—the maximum value of the effective sound level,
- LAF—the temporary value of the effective sound level,
- L_{CMPK}—the maximum peak sound level.

The measurements were taken during the morning rush hours (7–10 a.m.) in June 2021 at two types of locations:

1. hot-spots determined on the basis of an avifauna inventory—HP1 ÷ HP6,
2. spots adjacent to roads that are a source of traffic noise, i.e., S7 Express Road—Wybrzeże Gdyńskie Street from the east (R1-1 ÷ R1-6) and Podleśna Street from the south (R2-1 and R2-2).

3.2. The Surveys of the Avifauna

The previous inventories of avifauna in Warsaw were conducted during the breeding seasons and winter, in 1962–1975, 1988 and 2000 [44]. According to an inventory carried out in 2000, 42 species inhabited the Bielański Forest area (1–1.3 thousand pairs). The most numerous were starlings, blackbirds and chaffinches, whereas the most valuable for Warsaw were five species of woodpeckers, marsh tits, pied and red-breasted flycatchers. Over the period of 1988–2000, the total number of breeding species decreased slightly (from 44 to 42), though the diversity of the species composition has decreased significantly [44,45]. It is important to remark that the above-mentioned studies considered the avifauna of all of Warsaw and did not include detailed information on the bird distribution in Bielański Forest. The report contains the number of pairs per square meters (for common species), or the number of pairs (in the case of rare species or unevenly distributed ones). During the breeding seasons of 1992–1996 in the Bielański Forest Reserve, Mazgajski [46] analyzed changes in the numbers and nest sites of the great spotted woodpecker (*Dendrocopos major* Linnaeus, 1758) and the middle spotted woodpecker (*D. medius* Linnaeus, 1758). The population of the first species was relatively stable (10–14 pairs), with the latter being much more variable (1–2 pairs). The study indicated that the population of the analyzed species is related to the trees that woodpeckers choose as the sites of their nests. *D. major* prefers nesting in oaks, alders and pines, avoiding hornbeams. *D. medius* chooses predominantly oak. The positive effect of large, broadleaf trees on the bird population was, inter alia, confirmed by Kebrle et al. [47].

The presented studies on avifauna were focused on the area located within the boundaries of the NATURA 2000 site. A total of 18 surveys were carried out in May and June 2021, during the breeding season of the birds. Birds are most active during this period, usually marking their breeding territories by singing or its equivalent ("snares" of woodpeckers), and next building nests and raising their offspring. The observations were carried out at different times of the day: at dawn and in the morning, at daybreak, in the evening and at dusk, and also—though less frequently—in broad daylight, when it is easiest to observe woodpeckers feeding their young.

Due to the protection status of Bielański Forest, the classical transect penetration method [48] was not applicable. In the first stage of the research, the whole study area was inspected whilst being mindful of its environmental variability. At the same time, all of the points where rare species were observed in the entire study area were marked, as well as sites with a particularly high activity of various species. All of the points where individuals of selected species were observed (or heard) were also marked. After analysing the collected material, the highest density and diversity of avifauna (hot-spots) were determined. The six indicated avifauna hot-spots represent the different habitat types of the studied area, and were characterized by the highest bird activity. We also narrowed down the range of species on which we focused in further stages of the study. Five of them-the black woodpecker, green woodpecker, red-breasted flycatcher, pied flycatcher and redwing-were considered rare or very rare in Bielański Forest, and all of their sightings in the entire study area were recorded. In the case of the fourteen remaining species-cuckoo, great spotted woodpecker, wren, hedge-sparrow, robin, blackbird, fieldfare, song thrush, chiffchaff, willow warbler, great tit, blue tit, nuthatch and golden oriole—the focus was on hot-spots (a range of approximately 3000 sq. m., and also the sounds heard from the surrounding areas) and observations of breeding success, i.e., the number of nests found and young observed.

Furthermore, the location and characteristics of the individual hot-spots for avifauna were analysed with regard to the various zones of noise pollution that had been established in the previous stage of the study.

4. Results

The spatial changes of the acoustic climate in Bielański Forrest described by the W_{CA} index for time intervals (night—Lnight and daytime-evening-night—Lden) were assessed for two periods, i.e., 2007–2012 and 2012–2017. The results concerning the night-time noise level over the years 2007–2012 indicate some improvement of the situation. There was a significant increase in the area at the lowest noise level (<50 dB) and a slight decrease in the area exposed to noise levels above 60 dB. The situation in terms of $W_{CA\,Lnight}$ between the years of 2012 and 2017 was considered stable, although with a negative direction of changes. This is due to a decrease in the areas of low noise levels and an increase in areas exposed to high levels of noise. However, the differences in these changes are small and may be the result of the location of the measuring stations and the method of noise band interpolation. A similar situation occurred in the case of the $W_{CA\,Lden}$ index. The acoustic climate improved in the period 2007–2012, as the area of land with sound below 60 dB increased significantly. The area exposed to the highest noise (above 65 dB) also decreased. In the period of 2012–2017, the situation in terms of the Lden indicator deteriorated slightly. The areas exposed to noise levels below 60 dB decreased by about 6% in relation to the entire area of the site, whereas the areas that were exposed to noise levels of over 65 dB increased by about 4%. In the overall assessment, the acoustic climate should be considered stable, but with a slight negative trend (Table 1, Figure 2).

Taking into account the main scope of the study, including the assessment of noise pollution and the avifauna inhabiting the most environmentally valuable part of Bielański Forest, a similar quantitative analysis was carried out for the NATURA 2000 area. In the case of the $W_{CA\,Lden}$ index for the years 2007–2012, there was an increase in areas with a noise level below 60dB and a decrease in areas exposed to a noise level above 65 dB. At the same time, there was an extension of the area exposed to a noise level in the range of 60–65 dB. In the years 2012–2017, the situation was less favorable, because the extent of the areas affected by the highest noise levels increased slightly, while the area with the lowest sound level decreased to a medium extent.

Table 1. Changes in the areas exposed to road noise according to the Lnight and Lden indicators in 2007, 2012 and 2017 within Bielański Forest (our own elaboration on the basis of maps of noise pollution in Warsaw).

Indicator	Range of Noise Levels (dB)	Symbol	Area Exposed to Road Noise in the Individual Years (ha)			$W_{CA12-07}$ (%)	$W_{CA12-17}$ (%)
			2007	2012	2017		
Lnight	<50	$A_{Lnight<50}$	19.10	93.56	83.57	27.51	−3.70
	50–55	$A_{Lnight50-55}$	121.51	95.64	103.43	−9.56	2.89
	55–60	$A_{Lnight55-60}$	76.65	48.25	48.13	−10.49	−0.04
	60–65	$A_{Lnight60-65}$	31.41	20.03	17.74	−4.20	−0.85
	65–70	$A_{Lnight65-70}$	11.76	5.52	8.94	−2.31	1.27
	>70	$A_{Lnight>70}$	2.57	0.00	1.19	−0.95	0.44
Lden	<55	$A_{Lden<55}$	7.03	26.94	15.66	7.36	−4.18
	55–60	$A_{Lden55-60}$	64.18	118.92	110.92	20.23	−2.96
	60–65	$A_{Lden60-65}$	115.40	69.99	78.57	−16.78	3.18
	65–70	$A_{Lden65-70}$	51.07	36.43	36.90	−5.41	0.17
	70–75	$A_{Lden70-75}$	16.72	8.73	15.90	−2.95	2.66
	>75	$A_{Lden>75}$	8.60	1.99	5.05	−2.44	1.13

Grey cells indicate a negative trend.

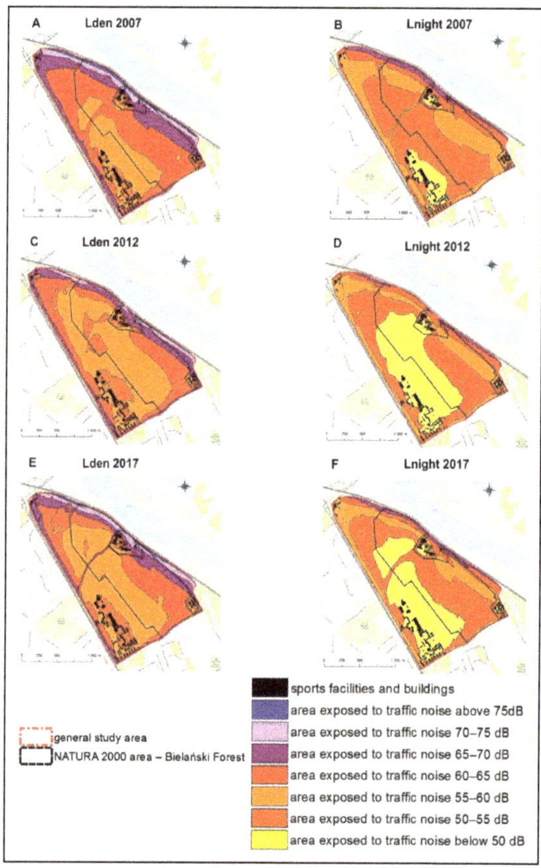

Figure 2. Changes in the areas exposed to road noise, as illustrated by the Lnight and Lden indicators in 2007 (**A,B**), 2012 (**C,D**) and 2017 (**E,F**) within the study area (our own elaboration on the basis of maps of noise pollution in Warsaw).

The analysis of the $W_{CA\,Lnight}$ index for the years 2007–2017 indicated minor changes with regard to the areas of high noise levels (around $\pm2\%$). Major changes, on the other hand, were observed in the case of the areas with the lowest noise level (<50 dB). In the period 2007–2012, the least noise-polluted area expanded by 35%, but in the years 2012–2017 it shrank by 10%. The reverse situation was observed in the case of the noise levels within the range of 50–60 dB. In the years of 2007–2012, the affected area decreased by 30%, but in the years 2012–2017, it increased again by about 10% (Table 2, Figure 2).

Table 2. Changes in the areas exposed to road noise according to the Lnight and Lden indicators in 2007, 2012 and 2017 in Bielański Forest NATURA 2000 (our own elaboration on the basis of maps of noise pollution in Warsaw).

Indicator	Range of Noise Levels (dB)	Symbol	Area Exposed to Road Noise in the Individual Years (ha)			$W_{CA12\text{--}07}$ (%)	$W_{CA17\text{--}12}$ (%)
			2007	2012	2017		
Lnight	<50	$A_{Lnight<50}$	0	45.40	32.42	34.97	−10.00
	50–55	$A_{Lnight50\text{--}55}$	69.36	53.58	65.46	−12.15	9.15
	55–60	$A_{Lnight55\text{--}60}$	43.95	21.45	22.31	−17.33	0.66
	60–65	$A_{Lnight60\text{--}65}$	11.04	8.42	5.80	−2.02	−2.02
	65–70	$A_{Lnight65\text{--}70}$	4.66	0.99	3.81	−2.83	2.17
	>70	$A_{Lnight>70}$	0.83	0	0.04	−0.64	0.03
Lden	<55	$A_{Lden<55}$	0	9.65	1.07	7.43	−6.61
	55–60	$A_{Lden55\text{--}60}$	25.24	71.34	61.23	35.51	−7.79
	60–65	$A_{Lden60\text{--}65}$	73.20	33.65	46.82	−30.46	10.14
	65–70	$A_{Lden65\text{--}70}$	22.70	12.99	13.92	−7.48	0.72
	70–75	$A_{Lden70\text{--}75}$	5.22	1.99	5.28	−2.49	2.53
	>75	$A_{Lden>75}$	3.48	0.22	1.52	−2.51	1.00

Grey cells indicate a negative trend.

The changes in the noise pollution during the analyzed period can also be illustrated by means of selected equal-sound pressure level contours (Figure 3). With regard to the previous and current maximum sound levels for recreational areas [12], spatial analyses were carried out for the following noise bands:

- 60 dB and 70 dB for Lden.
- 50 dB and 60 dB for Lnight.

It can be concluded that, in the case of Lden:

- the noise level of 70 dB was generated only from the side of Wybrzeże Gdyńskie Street; the course of the band in the years 2007–2017 was rather stable, and small changes occurred along the north-eastern border of the NATURA 2000 site;
- the course of the sound pressure level contour 60 dB varied in time; the most favorable situation was observed in 2012.

In the case of the noise level calculated for night time (Lnight), it must be stated that:

- Wybrzeże Gdyńskie Street was the main source of noise affecting the study area
- the course of the noise band 60dBA in the years 2012 and 2017 was rather stable and the changes in relation to the state as of 2007 amounted to a maximum of 50 m;
- the expansion of the area of the 50 dB noise band in the period 2012–2017 was also stable, especially in the northern and southern part of the NATURA 2000 site; in 2007, there were no areas exposed to noise levels below 50 dB.

Data concerning the noise pollution presented on the acoustic maps were supplemented with noise measurements taken in the selected locations within the studied area of the NATURA 2000 site (Table 3). The measurements of the noise levels represented by a temporary value of the effective sound level (the average sound value for a measurement) were performed for the hot-spots along the two main roads (R1-1 ÷ R1-6, R2-1 and R2-2) and also the hot-spots connected with areas hosting the most valuable species of birds (HP 1 ÷ HP 6). The conclusion suggested by the field study is that the obtained results are in line with the average noise values included in acoustic maps for 2017. The measurement locations are indicated in Figure 3.

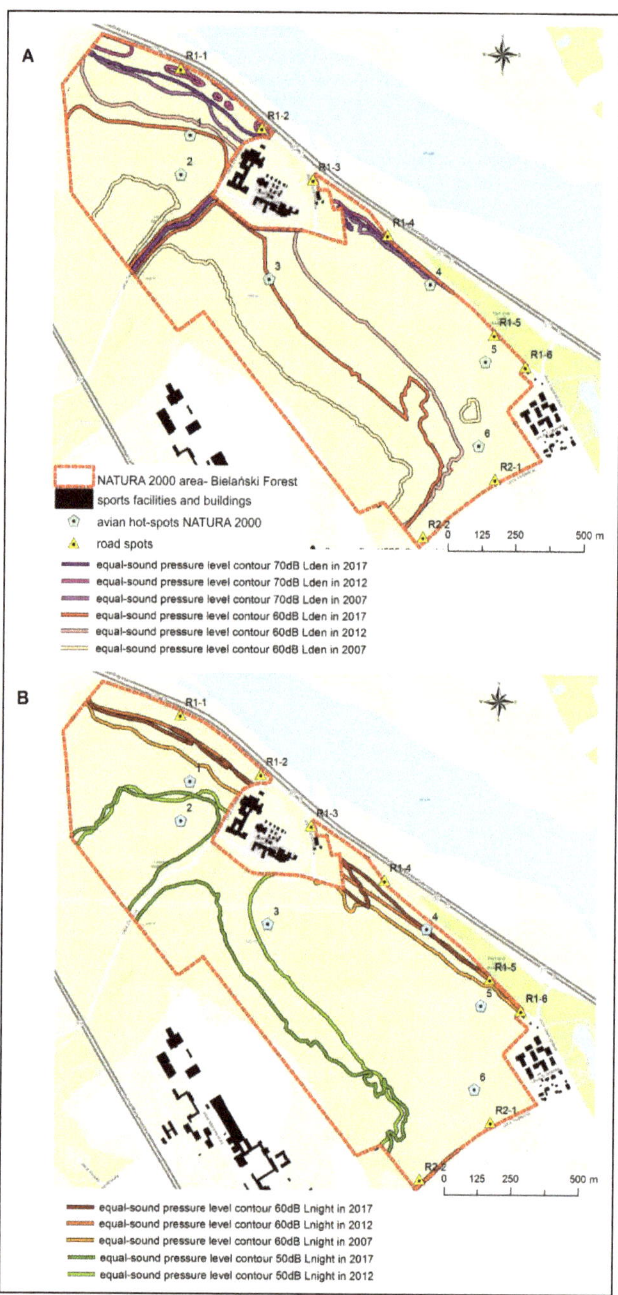

Figure 3. Changes in the location of the selected equal-sound pressure level contours in the years 2007, 2012 and 2017; (**A**)—for Lden, and (**B**)—for Lnight.

Table 3. The results of sound level measurements performed in June 2021 for avian hot-spots and road spots within the NATURA 2000 area (our own elaboration).

Main Groups of Measurement Spots	Measurement Spot Identifier [1]	$L_{AI\,min}$ (dB)	$L_{AI\,max}$ (dB)	L_{AF} (dB)	L_{CMPK} (dB)
Avian hot-spots	HP 1	49.5	70.8	59.7	86.9
	HP 2	43.8	62.6	52.6	84.2
	HP 3	45.2	67.9	47.8	92.5
	HP 4	47.4	113.3	63.3	138.1
	HP 5	48.6	88.5	54.0	115.3
	HP 6	41.5	72.0	65.7	89.9
Road spots for Wybrzeże Gdyńskie Street	R 1–1	63.4	84.0	65.5	94.1
	R 1–2	61.4	70.6	68.2	89.6
	R 1–3	63.9	84.5	66.5	98.2
	R 1–4	63.6	70.2	65.8	92.9
	R 1–5	55.5	94.2	69.0	121.3
	R 1–6	56.3	69.9	59.9	90.6
Road spots for Podleśna Street	R 2–1	49.1	69.9	56.6	93.5
	R 2–2	43.7	69.9	56.6	89.8

[1] The measurement spots are indicated in Figure 3.

As stated earlier, birds are a good indicator of progressive environmental changes, so they were considered a suitable group for studying the impact of noise on biocenosis. Species that are rare in Poland as a whole, and which are sporadically found in urban areas are considered to be the most important from the point of view of the natural attractiveness of the studied area. The distribution of the five bird species selected for the study, which were considered rare or very rare in Bielański Forest, is shown in Figure 4.

The black woodpecker (*Dryocopus martius* Linnaeus, 1758), considered to be one of the symbols of Bielański Forest, and a species which are rarely recorded during the breeding season within large cities -was encountered nine times: twice in the direct vicinity of avian hot-spots II, III and VI and, in other cases, in places overgrown with old trees with a large share of pine trees. The green woodpecker (*Picus viridis* Linnaeus, 1758) was observed seven times, mostly in the immediate vicinity of the Rudawka stream, in the alder or riparian forest. One of the observations was made in hot-spot IV. The least frequently observed species among those selected for the study was the red-breasted flycatcher (*Ficedula parva* Bechstein 1794), which was observed three times in the vicinity of hot-spots IV and V. The pied flycatcher (*Ficedula hypoleuca* Pallas 1764) was observed ten times, primarily within the alder or riparian forest near Rudawka stream, and in the vicinity of hot-spots III and V. The redwing (*Turdus iliacus* Linnaeus, 1758) was observed nine times, near hot-spots I, III and VI, and at several other points in the study area.

The study of the 14 species which were considered representative for Bielański Forest, and the most important from the point of view of the functioning of its biocenosis and attractiveness for nature lovers focused on hot-spots (see: Materials and Methods). The results of the observations are summarized in Table 4.

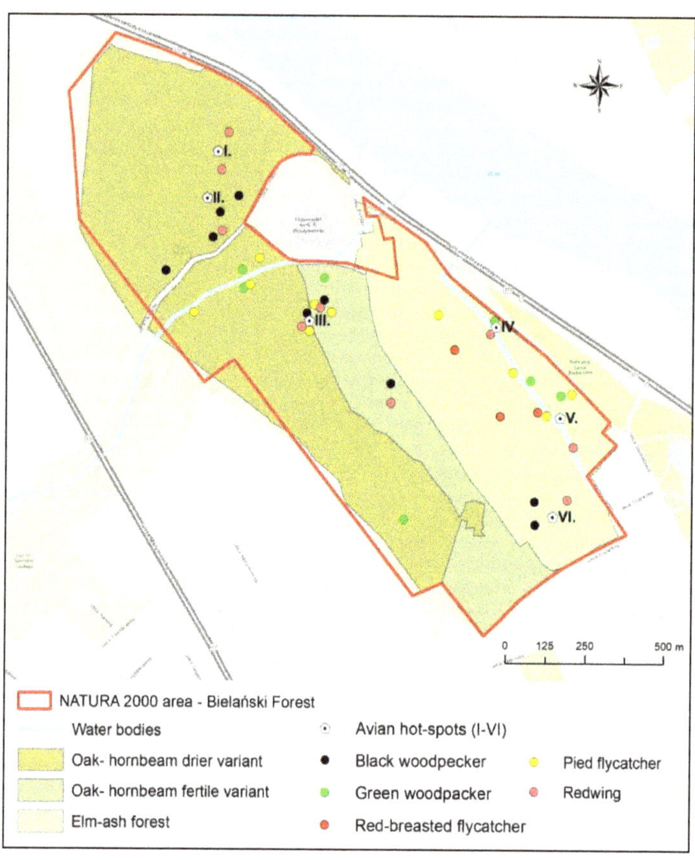

Figure 4. The locations of the five bird species considered particularly important for the biocenosis of Bielański Forest.

Table 4. The results of the bird observations at the hot-spots.

Species	Breeding Status [1]	Estimated Number of Individuals Per Avian Hot-Spot [2]	Remarks
Cuckoo *Cuculus canorus* Linnaeus, 1758	B	I-2; II-4; III-0; IV-0; V-5; VI-2	Does not build nests, breeding success dependent on other species to which it sends eggs
Great spotted woodpecker *Dendrocopos major* Linnaeus, 1758	B	I-6; II-15; III-6; IV-6; V-11; VI-2	In hot-spots I-V from 1 to 3 hollows with chicks
Wren *Troglodytes troglodytes* Linnaeus, 1758	US	I-0; II-4, III-2; IV-2;V-1; VI-0	
Hedge sparrow *Prunella modularis* Linnaeus, 1758	PB	I-4; II-0; III-3; IV-0, V-2; VI-2	
Robin *Erithacus rubecula* Linnaeus, 1758	B	I-3; II-8; III-1; IV-4; V-10; VI-2	In hot-spots II and V juveniles were observed shortly after leaving the nest

Table 4. *Cont.*

Species	Breeding Status [1]	Estimated Number of Individuals Per Avian Hot-Spot [2]	Remarks
Blackbird *Turdus merula* Linnaeus, 1758	B	I-9; II-1; III-3; IV- 16, V-2, VI-3	In hot-spot I the nest with 8 young was found, in hot-spot IV 2 nests with 5 and 6 young respectively
Fieldfare *Turdus pilaris* Linnaeus, 1758	B	I-2; II-8; III-11;IV-0; V-3; VI-9	In hot-spots II and VI nests with young were found, their number was estimated
Song thrush *Turdus philomelos* Brehm, 1831	B	I-8 *; II-2; III-6; IV-1; V-9; VI-2	* nest with 6 eggs, in hot-spot V nest with 6 young
Chiffchaff *Phylloscopus collybita* Boie, 1826	PB	I-1; II-3; III-2; IV-2; V-2; VI-0	
Willow warbler *Phylloscopus trochilus* Linnaeus, 1758	B	I-0; II-8; III-1; IV-2; V-2; VI-2	In hot-spot II the nest with 6 chicks
Great tit *Parus major* Linnaeus, 1758	B	I-10 *; II-2; III-20 *; IV-10 *; V-2; VI-10 *	* estimated abundance, young heard from occupied hollows
Blue tit *Cyanistes caeruleus* Linnaeus, 1758	B	I-0; II-10 *; III-2; IV-0; V-2; VI-10 *	* estimated abundance, young heard from occupied hollow and nest box
Nutchatch *Sitta europaea* Linnaeus, 1758	PB	I-2; II-0; III-2; IV-3; V-2; VI-1	
Golden oriole *Oriolus oriolus* Linnaeus, 1758	B	I-1; II-3; III-0; IV-2; V-5; VI-1	In hot-spot V 4 young shortly after leaving the nest

[1] B—breeding (nests found, adults observed with food in their beak or juveniles); PB—probably breeding (mating behavior observed, suitable nesting biotope); US—undetermined status (observations during feeding or voice recording); [2]—no. of hot-spot – no. of individuals; * indicates the additional explanation to the presented results, which acts as a supplement to better present the specific situation found in the peculiar site of the study area.

5. Discussion

The analysis of the acoustic maps for the years 2007–2017 (Figures 2 and 3, Table 2) and the results of the measurements carried out during the field work in 2021 (Table 3) confirmed that Wybrzeże Gdyńskie Street (checkpoints R1-1 ÷ R1-6) is the main source of noise pollution that affects Bielański Forest. Car traffic on Podleśna Street has a much smaller impact on the studied area. The impact of Marymoncka Street, which is located on the west of the Bielański Forest, is also insignificant. Road noise is variable over time and depends on many factors, such as: the road surface characteristics, the geometry and cross-section of the road, the traffic parameters and conditions, the speed of the vehicles, the type and shape of the surroundings, the weather conditions and others [49–51]. According to the measurements taken in the Wybrzeże Gdyńskie Street, there were 36,969 vehicles moving towards the city center and 37,743 in the opposite direction on an average day in 2012; in 2017, there were −47,647 and 48,082, and in 2019 there were −49,276 and 48,545 respectively [52]. However, in spite of the gradual growth in the daily traffic, the acoustic climate of the analysed area has remained quite stable. The slight improvement in the acoustic climate between 2007 and 2012 may be partly explained by the restoration or replacement of the road surface carried out in 2012 [53], whereas the deterioration of the acoustic conditions in 2017 may have been caused by the increase of the road traffic and a progressive decline in the technical state of the road. In 2020, some construction works took place that may also have resulted in a slight reduction of the noise pollution. According to the obtained results, the spatial range of high noise pollution was limited to the zone located along the main road. Negative changes were limited to a small area. This confirms the high noise-absorbing ability of the existing vegetation. It is also in line with the studies regarding the relationship between land cover and noise reduction [7–10].

As has already been stated in the initial part of the study, the negative impact of urban noise on birds has been reported by other scientists [23–28,54]. However, the responses to noise can vary between species and the biotope in which a particular avifauna assemblage occurs [55]. The results of the study for five species (Figure 4) which were considered the most valuable in the Bielański Forest area are inconclusive. One of the birds—the red-breasted flycatcher—was observed so sporadically that no firm conclusions could be drawn in its case. It is worth mentioning, however, that the species in question—very few in the scale of the whole country—have been present in Bielański Forest since at least 1988 [44]. It seems symptomatic, however, that the locations of its sightings were wet meadows and alder woodlands, which are consistent with the general habitat preferences of this species [56], even though they are located in a part of the reserve with a strong acoustic impact from the expressway. This could suggest that, for this species, habitat requirements take precedence over possible adverse noise impacts. A similar trend can be observed in the case of the pied flycatcher (which prefers dense deciduous forests, especially oak), while the black woodpecker seems to avoid areas with the highest intensity of acoustic stimuli, although according to its biology, it prefers old coniferous forests, especially with pine and spruce. This is a tendency which has been observed for years with regard to this species in the study area [46]. In the case of the green woodpecker and redwing, it is difficult to draw any scientifically justified conclusions. The analysis of the occurrence of the 14 species which were found to be the most important from the point of view of the functioning of the whole biocenosis is more interesting, seeing as their abundance is high enough to draw more far-reaching conclusions (Table 4).s

The conducted research—apart from the analysis of the impact of the noise intensity on the occurrence of key species of avifauna—made it possible to determine the trends in the settlement of the research area by birds in the last 20 years. In 2001, a monograph devoted to the birds of Warsaw [44] was published with detailed information on the abundance and occurrence of individual species within the city. The comparison of the data contained therein with the results of observations from 2021 (Table 4, Figure 4) is a valuable material illustrating the trends in the settlement/withdrawal of various species from areas subject to increasing anthropopressure. In the case of five species—the fieldfare, nuthatch, chiffchaff, willow warbler and song thrush—no significant changes in terms of their abundance in the area under study were found. The aforementioned birds seem to form stable populations in urban green areas, and from year to year, they became a typical element of Warsaw's avifauna. A very interesting tendency was noted in the case of the redwing—the first breeding of this species in Bielański Forest was observed only at the end of the 20th century [44]. The results of observations from 2021 indicate that it now certainly nests in the studied area, and its population has increased significantly. A minor, though noticeably similar, tendency was also noticed in the case of the cuckoo, black woodpecker (in 2001, there was probably only one breeding pair in the Bielański Forest), and wren. The green woodpecker, for which a downward trend in abundance (two breeding pairs) was shown in the late 20th century, certainly maintains this abundance status and, in all likelihood, is more numerous. A similar situation may be stated for the great spotted woodpecker. Mazgajski [46] writes that, from the 1960s to the end of the 20th century, a population size of 10–14 pairs was maintained; this can be assessed as being higher in 2021. In the case of woodpeckers, an important element that may affect their occurrence in the studied area is the maturation of the stand and a greater number of available hollows (the green woodpecker does not forge them on its own, but uses already existing ones) and trees suitable for woodpeckers which create their own (in the case of the other two species). In the case of the hedge sparrow, the greatest obstacle to the development of urban populations was identified by Luniak et al. [44] as the decrease in suitable breeding biotopes. It seems that this problem does not occur in Bielański Forest, hence the population of this species remains more or less stable. The great tit is the most common of the investigated species, with its population appearing to be stable. A similar situation occurs in the case of the golden oriole;—Bielański Forest was and still is one of the sites with the biggest population of this species within the administrative borders of Warsaw. The population of the blue tit also seems to be stable, although in the case of this species, there is a clear correlation between the numbers in specific zones of the problem area and the number of available nest boxes. This is an aspect which is often highlighted in the literature [44]. Robins and blackbirds are fairly common species throughout the study area, which is in agreement with the status 20 years ago, and indicates that the populations of both of the mentioned species are stable.

The analysis of the data concerning the noise level (Table 3) would suggest that hot-spot IV, with the highest maximum value of the effective sound level, with a peak of 138.1 dB, will be avoided by birds. Indeed, five species out of the 14 surveyed were absent, i.e., —the blue tit, chiffchaff, fieldfare, hedge sparrow and cuckoo. However, the remaining nine were present, and species such as the great spotted woodpecker, blackbird, green woodpecker and red-breasted flycatcher were found in large numbers within or adjacent to the hot-spot. The differences in the species composition between the remaining hot-spots were small, although more nests were found in some than in the others (e.g., hot-spot II, with three great spotted woodpecker hollows).

The analysis of the noise distribution seems to indicate that the factor determining the distribution and abundance of the key species in the study area is the type of habitat, rather than the differences in noise levels, which appear to be less important. The last above-mentioned conclusion emphasizes the importance of maintaining the habitats preferred by birds. In the case of Bielański Forest, the ground water level is the key factor determining the quality and occurrence of elm-ash forests and oak-hornbeam forests. However, the water conditions in the forest depend not only on land-use in this area but also in its surroundings. Recommendations concerning land-use planning include actions that are also set up in the current Protection Plan of Bielański Forest Reserve and NATURA 2000 [57], and other related publications [5,34,36]. In the case of areas located in the western and eastern vicinity of Bielański Forest, it is important to limit new developments and introduce specific solutions for sustainable water management that increase bioretention and small-scale water retention. It is also important to maintain the ecological connectivity of the area, especially in the north-western direction toward Młociński Forest and onwards to Kampinoski Forest, inter alia along the area of the NATURA 2000 site PLB140004 PLB1—The Middle Vistula River Valley. This objective also requires the maintenance of existing green areas and introduction of various new types of green areas, including forests and recreational areas.

6. Conclusions

The final conclusions are as follows:

1. Establishing NATURA 2000 areas in cities is justified by the necessity of protecting and maintaining their biodiversity, including valuable and unique plant and animal species, as well as biotops occurring in the given area. At the same time, the natural protection of the given areas is sometimes of great significance for the functioning of the environmental system of the entire city. This pertains not only to the biological subsystem, but also to the climatic and hydrological subsystems. These areas are also used for recreation. The NATURA 2000 area of Bielański Forest is an example of an object with multifunctional significance. The condition of its natural environment is dependent on local and supralocal factors.

2. The analyzed area is located in the direct proximity of the main city roads, and is within the impact zone of increased noise levels. The analysis of the changes in the noise pollution carried out for the years 2007–2017 indicates that despite the gradually increasing traffic, the negative changes in the scope and intensity of the noise are not significant. This results mainly from the existing vegetation cover. Dense

and deciduous forests significantly reduce noise pollution, even at a relatively short distance from roads.

3. The inventory of the avifauna in the area of Bielański Forest confirms the occurrence of many valuable species. The state of the population of the majority of the analyzed species is relatively stable. The key bird species of the NATURA 2000 study area seem to have adapted to urban conditions. A strong correlation between the noise distribution and certain species was not observed.

4. Habitat type is the determining factor for the distribution and abundance of keystone species in the study area—noise levels seem to be less important. Therefore, in the case of Bielański Forest, it is necessary to focus on actions that guarantee the maintenance of the existing status and counteract the deterioration of habitats, especially those of the riparian forests.

Author Contributions: Conceptualization, A.P.-Z., K.P.; Formal analysis, A.P.-Z., K.P.; Investigation, A.P.-Z., P.O., K.P.; Methodology, A.P.-Z., P.O., K.P.; Project administration, A.P.-Z.; Software, A.P.-Z., K.P.; Supervision, A.P.-Z.; Validation, A.P.-Z., P.O.; Visualization, A.P.-Z., K.P.; Writing—original draft, A.P.-Z., P.O.; Writing—review and editing, A.P.-Z. All authors have read and agreed to the published version of the manuscript.

Funding: Publication financed by the "Excellent Science" Program of the Minister of Science and Higher Education DNK/SP/466229/2020.

Conflicts of Interest: The authors declare no conflict of interest.

References

1. Sundseth, K.; Raeymaekers, G. Biodiversity and natura 2000 in Urban Areas. In *Nature in Cities across Europe: A Review of Key Issued and Experiences*; IBGE/ BIM: Brussels, Belgium, 2006; p. 89.

2. Szulczewska, B.; Kaftan, J. *Kształtowanie Systemu Przyrodniczego Miasta [Planning of the Natural Areas System of the City]*; IGPiK: Warsaw, Poland, 1996.

3. Kaliszuk, E. Funkcje systemu przyrodniczego miasta w kształtowaniu warunków środowiska przyrodniczego na przykładzie Warszawy. *Prace i Studia Geograficzne* **2005**, *36*, 35–47.

4. Skorupski, J.; Falkowski, M.; Gnyś, E.; Talarek, R.; Ostaszewska, E. *Analiza obszarów NATURA 2000 pod kątem ich potencjalnych zagrożeń wynikających z lokalizacji nowych przedsięwzięć na terenie m. st. Warszawy*; Biuro Planowania Rozwoju Warszawy S.A.: Warsaw, Poland, 2009; p. 24.

5. KIPPiM. *Atlas Ekofizjograficzny Miasta Stołecznego Warszawy*; Krajowy Instytut Polityki Przestrzennej i Mieszkalnictwa: Warsaw, Poland, 2018.

6. Kaymaz, I.; Akpınar, N.; Belkayalı, N. *User Persception of Soundscapes of Urban Parks in Ankara*; METU: Ankara, Turkey, 2012.

7. Margaritis, E.; Kang, J. Relationship between Green Space-Related Morphology and Noise Pollution. *Ecol. Indic.* **2017**, *72*, 921–933. [CrossRef]

8. Zimny, H. *Ochrona i Kształtowanie Zieleni w Aglomeracjach Miejskich (Protection and Shaping of Greenery in Urban Agglomerations)*; Zarząd Główny LOP: Warsaw, Poland, 1978.

9. Samara, T.; Tsitsoni, T. The Effects of Vegetation on Reducing Traffic Noise from a City Ring Road. *Noise Control. Eng. J.* **2011**, *59*, 68–74. [CrossRef]

10. Van Renterghem, T. Towards Explaining the Positive Effect of Vegetation on the Perception of Environmental Noise. *Urban For. Urban Green.* **2019**, *40*, 133–144. [CrossRef]

11. Berglund, B.; Lindvall, T.; Schwela, D.H. *Guidelines for Community Noise*; WHO World Health Organization: Geneva, Switzerland, 1999.

12. Obwieszczenie Ministra Środowiska z Dnia 15 Października 2013 r. w Sprawie Ogłoszenia Jednolitego Tekstu Rozporządzenia Ministra Środowiska w Sprawie Dopuszczalnych Poziomów Hałasu w Środowisku (Ordinance of the Minister of the Environment of 15 October 2013 on the Permissible Noise Levels in the Environment). Dz. U. 2014 poz. 112. Available online: http://isap.sejm.gov.pl/isap.nsf/DocDetails.xsp?id=WDU20140000112 (accessed on 18 May 2021).

13. Brown, A.L. Acoustic Objectives for Designed or Managed Soundscapes. *Soundscape J. Acoust. Ecol.* **2003**, *4*, 19–23.

14. Payne, S.R. Urban park soundscapes and their perceived restorativeness. *Proc. Inst. Acoust. Belg. Acoust. Soc* **2010**, *32*, 264–271.

15. Tse, M.S.; Chau, C.K.; Choy, Y.S.; Tsui, W.K.; Chan, C.N.; Tang, S.K. Perception of Urban Park Soundscape. *J. Acoust. Soc. Am.* **2012**, *131*, 2762–2771. [CrossRef]

16. Jaszczak, A.; Małkowska, N.; Kristianova, K.; Bernat, S.; Pochodyła, E. Evaluation of Soundscapes in Urban Parks in Olsztyn (Poland) for Improvement of Landscape Design and Management. *Land* **2021**, *10*, 66. [CrossRef]

17. Brlík, V.; Šilarová, E.; Škorpilová, J.; Alonso, H.; Anton, M.; Aunins, A.; Benkö, Z.; Biver, G.; Busch, M.; Chodkiewicz, T.; et al. Long-Term and Large-Scale Multispecies Dataset Tracking Population Changes of Common European Breeding Birds. *Sci. Data* **2021**, *8*, 21. [CrossRef]

18. Grimm, N.; Faeth, S.; Golubiewski, N.; Redman, C.; Wu, J.; Bai, X.; Briggs, J. Global Change and the Ecology of Cities. *Science* **2008**, *319*, 756–760. [CrossRef] [PubMed]

19. Chiesura, A. The Role of Urban Parks for the Sustainable City. *Landsc. Urban Plan.* **2004**, *68*, 129–138. [CrossRef]

20. Kinzig, A.P.; Warren, P.; Martin, C.; Hope, D.; Katti, M. The Effects of Human Socioeconomic Status and Cultural Characteristics on Urban Patterns of Biodiversity. *Ecol. and Soc.* **2005**, *10*, 23. [CrossRef]

21. Aronson, M.F.J.; La Sorte, F.A.; Nilon, C.H.; Katti, M.; Goddard, M.A.; Lepczyk, C.A.; Warren, P.S.; Williams, N.S.G.; Cilliers, S.; Clarkson, B.; et al. A Global Analysis of the Impacts of Urbanization on Bird and Plant Diversity Reveals Key Anthropogenic Drivers. *Proc. R. Soc. B Biol. Sci.* **2014**, *281*, 20133330. [CrossRef] [PubMed]

22. McMahon, B.J.; Doyle, S.; Gray, A.; Kelly, S.B.A.; Redpath, S.M. European Bird Declines: Do We Need to Rethink Approaches to the Management of Abundant Generalist Predators? *J. Appl. Ecol.* **2020**, *57*, 1885–1890. [CrossRef]

23. Marzluff, J.M. Worldwide urbanization and its effects on birds. In *Avian Ecology and Conservation in an Urbanizing World*; Marzluff, J.M., Bowman, R., Donnelly, R., Eds.; Springer: Boston, MA, USA, 2001; pp. 19–47. ISBN 978-1-4615-1531-9.

24. Kleist, N.J.; Guralnick, R.P.; Cruz, A.; Francis, C.D. Anthropogenic Noise Weakens Territorial Response to Intruder's Songs. *Ecosphere* **2016**, *7*, 11. [CrossRef]

25. Seress, G.; Liker, A. Habitat Urbanization and Its Effects on Birds. *Acta Zool. Acad. Sci. Hung.* **2015**, *61*, 373–408. [CrossRef]

26. Halfwerk, W.; Holleman, L.J.M.; Lessells, C.M.; Slabbekoorn, H. Negative Impact of Traffic Noise on Avian Reproductive Success. *J. Appl. Ecol.* **2011**, *48*, 210–219. [CrossRef]

27. Proppe, D.S.; Sturdy, C.B.; Clair, C.C.S. Anthropogenic Noise Decreases Urban Songbird Diversity and May Contribute to Homogenization. *Glob. Chang. Biol.* **2013**, *19*, 1075–1084. [CrossRef]

28. Manzanares, M.L.; Macías, G.C. Songbird Community Structure Changes with Noise in an Urban Reserve. *J. Urban Ecol.* **2018**, 1–8. [CrossRef]

29. Hanna, D.; Blouin-Demers, G.; Wilson, D.R.; Mennill, D.J. Anthropogenic Noise Affects Song Structure in Red-Winged Blackbirds (*Agelaius phoeniceus*). *J. Exp. Biol.* **2011**, *214*, 3549–3556. [CrossRef]

30. PLH140041 Natura 2000 Obszar Natura 2000. Available online: http://crfop.gdos.gov.pl/CRFOP/widok/viewnatura2000.jsf?fop=PL.ZIPOP.1393.N2K.PLH140041.H (accessed on 12 July 2021).

31. *Studium Uwarunkowań i Kierunków Zagospodarowania Przestrzennego m.st. Warszawa (Study of conditions and directions of spatial development of the city of Warsaw)*; Warsaw Architecture and Spatial Planning Department. Warsaw Municipality: Warsaw, Poland, 2006; (amended 2010, 2014, 2021. Available online: https://bip.warszawa.pl/NR/exeres/CC1A6257-AED9-4E29-827C-EFE38C9 60FD8,frameless.htm (accessed on 10 June 2021).

32. Local Data Bank. Available online: https://bdl.stat.gov.pl/BDL/start (accessed on 28 June 2021).

33. Pawłat, H.; Wanke, A. *Raport Oddziaływania Wody w Potoku Rudawka oraz Przerzutu Wody z Potoku RUDAWKA do Zbiornika Kępa Potocka*; Biuro Konsultacyjne "Inżynieria Środowiska": Warsaw, Poland, 2001.

34. Fornal-Pieniak, B.; Żarska, B.; Zaraś, E. Natural Evaluation of Landscape in Urban Area Comprising Bielański Forest Nature Reserve and Surroundings, Warsaw, Poland. Directions for Landscape Protection and Planning. *Ann. Wars. Univ. Life Sci. Land Reclam.* **2018**, *50*, 327–338. [CrossRef]

35. Podawca, K.; Krzysztof, K.; Pawłat-Zawrzykraj, A. The Assessment of the Suburbanisation Degree of Warsaw Functional Area Using Changes of the Land Development Structure. *Misc. Geographica. Reg. Stud. Dev.* **2019**, *23*, 215–224. [CrossRef]

36. Program ochrony środowiska dla m.st. Warszawy na lata 2021–2024 (Environmental Protection Programme for the Capital City of Warsaw for 2021–2024). Available online: https://bip.warszawa.pl/NR/exeres/CA8EF8BF-F40C-4511-8C7A-1EBF054EE365.htm (accessed on 12 July 2021).

37. Podawca, K.; Karpiński, A. Analysis of Spatial Development Possibilities of Properties Endangered by Road Noise in the Context of Permissible L N and L DWN Indicators. *J. Ecol. Eng.* **2021**, *22*, 238–248. [CrossRef]

38. Podawca, K.; Staniszewski, R. Impact of Changes of the Permissible Railway Noise Levels on Possibilities of Spatial Management in Urban Areas. *Rocznik Ochrona Środowiska* **2019**, *21*, 1378–1392.

39. Podawca, K.; Karsznia, K.; Jewuła, K. Analysing the Impact of Modified Railways Noise Thresholds in Poland on the Development of Residential Areas. *Misc. Geogr.* **2021**, *25*, 155–168. [CrossRef]

40. Mapy Akustyczne m.St. Warszawy (Acoustic Map of Warsaw). Available online: http://mapaakustyczna.um.warszawa.pl/pl/mapa/mapa20xx.html (accessed on 28 June 2021).

41. Hutchinson, M.F. Adding the Z Dimension. In *The Handbook of Geographic Information Science*; John Wiley & Sons: Hoboken, NJ, USA, 2008; pp. 144–168. ISBN 978-0-470-69081-9.

42. Hutchinson, M.F.; Gallant, J.C. Digital Elevation Models and Representation of Terrain Shape. In *Terrain Analysis, Principles and Applications*; Wilsons, J.P., Gallant, J.C., Eds.; John Wiley and Sons: New York, NY, USA, 2000; pp. 29–49.

43. *Wytyczne Do Sporządzania Map Akustycznych (The Guidelines for Noise Maps)*; Instytut Ochrony Środowiska: Warsaw, Poland, 2006. Available online: https://www.google.com/url?sa=t&rct=j&q=&esrc=s&source=web&cd=&ved=2ahUKEwj3zLGq-ZvzAhWRzosKHeXQD6lQFnoECAcQAQ&url=http%3A%2F%2Fwww.gios.gov.pl%2Fimages%2Fdokumenty%2Fsprawozdawanie%2FWytyczne_do_sporzadzania_map_akustycznych_2016.pdf&usg=AOvVaw1HGh-YbCmanOzXv758itad (accessed on 5 June 2021).
44. Luniak, M.; Kozłowski, P.; Nowicki, W.; Plit, J. *Ptaki Warszawy 1962–2000 (Birds in Warsaw)*; IGiPZ PAN: Warsaw, Poland, 2001.
45. Chojnacki, J.M.; Czajka, L.; Luniak, M.; Miścicki, S.; Namura-Ochalska, A. Raport o wartości i stanie przyrody Lasu Bielańskiego (The report on the value and state of Nature of the Bielański Forest). Available online: http://lasbielanski.waw.pl/cel-i-charakter-raportu.html (accessed on 12 July 2021).
46. Mazgajski, T. Changes in the numbers and nest sites of the Great Spotted Woodpecker Dendrocopos major and Middle Spotted Woodpecker D. medius in the Las Bielański Reserve in Warsaw. *Ochr. Przyr.* **1997**, *54*, 155–160.
47. Kebrle, D.; Zasadil, P.; Hošek, J.; Barták, V.; Astný, K. Large Trees as a Key Factor for Bird Diversity in Spruce-Dominated Production Forests: Implications for Conservation Management. *For. Ecol. Manag.* **2021**, *496*, 119460. [CrossRef]
48. *Monitoring Ptaków Lęgowych Poradnik Metodyczny (Monitorin of Breeding Birds. The Methodological Guidebook)*, 2nd ed.; Chylarecki, P.; Sikora, A.; Cenian, Z.; Chodkiewicz, T. (Eds.) GIOŚ: Warsaw, Poland, 2015; ISBN 978-83-61227-45-8.
49. Gierasimiuk, P.; Motylewicz, M. Hałas w otoczeniu dróg i ulic-problemy oceny i działania ochronne (Traffic noise in the vicinity of roads and streets—Assassment problems and protective actions). In *Inżynieria Środowisk—Młodym Okiem*; Skoczko, I., Piekutin, J., Zarzecka, A., Gregorczuk-Petersons, E., Eds.; Oficyna Wydawnicza Politechniki Białostockiej: Białystok, Poland, 2014; pp. 59–93. ISBN 978-83-62582-53-2.
50. Sandberg, U. Road Traffic Noise—The Influence of the Road Surface and Its Characterization. *Appl. Acoust.* **1987**, *21*, 97–118. [CrossRef]
51. De León, G.; Pizzo, A.D.; Teti, L.; Moro, A.; Bianco, F.; Fredianelli, L.; Licitra, G. Evaluation of Tyre/Road Noise and Texture Interaction on Rubberised and Conventional Pavements Using CPX and Profiling Measurements. *Road Mater. Pavement Des.* **2020**, *21*, S91–S102. [CrossRef]
52. Analiza ruchu na drogach (Analysis of Road Traffic). Available online: https://zdm.waw.pl/dzialania/badania-i-analizy/analiza-ruchu-na-drogach/ (accessed on 5 July 2021).
53. ZDM Trzy frezowania w ten weekend: Wybrzeże Gdyńskie, Kijowska i al. Niepodległości. Available online: https://warszawa.naszemiasto.pl/trzy-frezowania-w-ten-weekend-wybrzeze-gdynskie-kijowska-i/ar/c4-1398171 (accessed on 5 July 2021).
54. Dorado-Correa, A.M.; Rodríguez-Rocha, M.; Brumm, H. Anthropogenic Noise, but Not Artificial Light Levels Predicts Song Behavior in an Equatorial Bird. *R. Soc. Open Sci.* **2016**, *3*, 160–231. [CrossRef]
55. Nemeth, E.; Brumm, H. Birds and Anthropogenic Noise: Are Urban Songs Adaptive? *Am. Nat.* **2010**, *176*, 465–475. [CrossRef] [PubMed]
56. Svensson, L.; Mullarney, K.; Zetterstrom, D. *Collins Bird Guide*, 2nd ed.; Harpers Collins Publishers: London, UK, 2009.
57. Plan Ochrony Rezerwatu Las Bielański i Obszaru Natura 2000 PLH 140041 "Las Bielański". Zarządzenie RDOŚ w Warszawie z Dnia 22 Września 2016 r. Dz.Urz. Woj. Mazowieckiego, 26 Poz. 8575 [Protection Plan of Nature Reserve Bielański Forest and the Nature 2000 Area PLH 140041 "Bielański Forest". Regulation of Regional Director for Environmental Protection of September 22, 2016. Official Journal of Mazowieckie Province 26, Position 8575]. Available online: http://bip.warszawa.rdos.gov.pl/zarzadzenie-regionalnego-dyrektora-ochrony-srodowiska-w-warszawie-z-dnia-22-wrzesnia-2016-r-publikowane-w-dzienniku-urzedowym-wojewodztwa-mazowieckiego-poz-8575 (accessed on 5 June 2021).

MDPI

St. Alban-Anlage 66

4052 Basel

Switzerland

Tel. +41 61 683 77 34

Fax +41 61 302 89 18

www.mdpi.com

Forests Editorial Office

E-mail: forests@mdpi.com

www.mdpi.com/journal/forests